The Heart of Reality

It is an emotional Heart That Propels us into the Future

AF808400

E. R. Margis

INFINITY PUBLISHING

ISBN 978-1-4958-0065-8
ISBN 978-1-4958-0066-5 eBook

Printed in the United States of America

Published March 2014

INFINITY PUBLISHING
1094 New DeHaven Street, Suite 100
West Conshohocken, PA 19428-2713
Toll-free (877) BUY BOOK
Local Phone (610) 941-9999
Fax (610) 941-9959
Info@buybooksontheweb.com
www.buybooksontheweb.com

Warning and Disclaimer

On-going efforts were made during the writing and before publication of "The Heart of Reality" to make it's content accurate and free of errors. However there may be mistakes both typographical and in content.

A new view of the universe is taken in this work, please be advised if you zip through the text by skimming you may miss some interesting content. The Universe is a great mystery, and this book can be read like a mystery story, a piece at a time.

The Heart of Reality, of the book's title

Love, Wisdom, and Justice, are important needs in a person's life, but as many have found out, there is more to life. **"The Heart of Reality"** experiences emotion and, **modifies the flow of change** within us, and within the natural world,

The Illustration of The Heart Of Reality on this book's cover

The palm tree of the cover illustration represents, "The Inevitable", the **"Cycle of life",** of all born or made things. The date palm (Phoenix dactylifera), of which there are some thirty species, is widely known from antiquity in parts of Asia, Pakistan, India, Egypt, and northern Africa, where some refer to it as **"The tree of life"**, it can have a life span as long as 150 years, and is reborn in it's seedlings, a form of eternal life, much as other living things, are reborn in their children. Date palms are of male or female gender. Commercial growers keep a few male trees around to fertilize the blossoms of the many female trees (which bear the fruit, the dates). The date palm is subjected to the vagaries of life, such as, **chance** experiences, and the nutrients it **selects** to use from the air and soil. Chance is represented in the illustration by two dice, Choice or Selection, is represented by pieces of a picture puzzle. The wedge is "discrimination", separating non-complementary opposites. The "It", of the cover text is, **The Heart of Reality,** "The Controller of Controllers".

Table of Contents

Part One (of three parts)
The Eternal Cosmic Ocean

Part Two
The Heart of Reality

The Doggedly Curious, and Science

Electromagnetic Radiation, Wave or Particle, What is "Matter?

Part Three

The Self

Matrix diagrams

Notes:

Note: The term **man** in this book means **a person of either gender**.

If you are interested in wave motion of all kinds in different media, an excellent book on the subject is:

"The Wave Watcher's Companion", by Gavin Pretor - Pinney
Copyright 2010, A Perigee trade paperback. ISBN: 978-0-399-53426-3.

Notes continued

Notes continued

About Writing this book?

This book was written to give an improved understanding of nature, not to torment higher form school boys or young people in love.
When I began looking into the covered subjects, to my mind, current scientific literature left key questions unanswered, gaps in understanding like missing pieces in nature's jig saw puzzle, I have tried to fill in a few of those pieces by showing their relationships to each other.
So as to not forget, I made notes here and there along the way of discovery, and eventually, for better or worse, wrote down my reasoned conclusions. An Australian Aborigine might say of the result, this book, is a story that explains, explains the ways of nature and a person's place in it, it is a story of the seen and unseen, and when you read it, it is a "**Dreamtime**".

An East Indian might see learning as "The Atman", "The Self", experiencing life, the self of each living thing and "The Self", of the universe as a whole.

A discovery-learning process, is an adventure, not accomplished using a time clock, or with someone saying "We do not want it good, we want it Friday." I believe I have made a start at a different view of nature, by presenting, a theory, along with matrix diagrams showing relationships of terms in those diagrams, and have added **a glossary** of useful definitions of words found in the text. There

is also **an index** of page numbers where some of the things mentioned in the text can be found.
If you embark on a similar quest, consider reading a little, thinking a little, and relax, after all how difficult can it be to write. If you are French maybe have some wine and cheese. If you are German, remember "dreams are not done by the numbers."

This quest to understand was driven by **"a need to know"**, as are the quests of many others. What child in grade school did not wonder what, white school paste, tasted like, I venture many gave it a lick, I did. The bank robber lying on the pavement in the movie **Dirty Harry**, said "I gots to know", he wanted to know if the gun Dirty Harry (played by actor Clint Eastwood) held him at bay with was still loaded. Dirty Harry claimed, it was after all "the most powerful handgun in the world, and it will blow your head clean off".* Curiosity drove the thinkers of long ago to sort out their world as best they could, by asking "what made the winds"? "What power holds a person against the face of the Earth"? "What is Life"? And asking a ton of other questions, including perhaps as they looked up on a starry night, "Where did this all come from"? In answer, some spoke of "Levels of being", "Planes of Existence", or "The Aether", A few spoke of a great unity, a vast ocean, an eternal **"Cosmic Ocean**."

Note: The gun used by "Dirty" Harry Callahan in the 1971 Movie was a **.**44 Magnum, Smith and Wesson M29 revolver. The bullet it fired had a diameter of approximately 10.9mm (a bit over 4/10'ths of an inch). The propellant charge of a magnum bullet can be greater than that of an ordinary bullet of the same caliber, the gun itself is built stronger to withstand the greater propelling forces.

Over the centuries many persons have wondered just what controls their destiny, in other words, what is **the regulating, controlling, principle** in their lives, and what determines the destiny of atoms, plants, suns, moons, and gods, yes even the gods. Some have personified the regulating "means", I have named them "The Heart of Reality", and if it skips a beat now and then, the human heart can be like that also. The pulse of "The Heart of Reality", is the pulse of life.

This control center attentively adjusts the interaction of complementary opposites in the universe, opposites such as Potentiality and Actuality, Substance and Non-Substance. Between all opposites there is a "tension". Communicating "**Forces**" link the opposites, Effort forces attempt to bring change, "**Counter forces",** resist efforts to bring change.

The regulating "means", controlling the Heart of Reality's pulse, determines, the characteristics, behaviors, and shapes of all things in the universe including those of each individual, be it a bird, bee, butterfly or man.

In this present work I have looked anew at definitions of some words and concepts found in the text, I have reevaluated and improved them as needed, the reworked definitions can be found in this book's glossary, they take into account traditional meaning's, plus the influence of my recent work on their relationships to one another.

The relationships of basic terms are shown in this book using a simple tic-tac-toe style graphic,

"Matrix Diagram". There are Individual "Matrix Diagrams", for "The Cosmic Ocean", "The Self", "The Mind" and one for the mythic "Chinese Dragons". Each diagram shows basic terms and their relationship to others in the diagram.
In my previous works **"The Man Who Saw Space and Time"**, and **"I Am Universe"**, some of man's previous views of the natural world were revealed, including views on the nature of the "Self".

As with my other books, This one, is not on politics, religion, or what sport team is better, also the explanations, definitions and other information in this book may contain errors both typographical and in content ,even though on-going efforts were made during writing to make the text accurate.

About the words used in this book

One advantage of using and writing in the English language is that it offers "a ton of words" to choose from, a great diversity, with which a person can express their thoughts, ideas, discoveries etc. In this book I have tried to use commonly used words to be better understood by many persons, if you find a word, and are unsure of it's meaning, please, look it up in **the glossary** near the end of this book, there you will find the meaning usually intended in the text by the author.
If you have better explanations or word definition than those in the glossary, and this book is yours, write in your notations, or type them out on a piece of paper and paste it in along the binding edge of the book on an appropriate page, a side effect of this practice may be to ruin the resale value of the

book, however a better understanding of the subject may offset this disadvantage.

As an aside, let me stray two ticks from the subject of this book, to give my feelings about books in general.
I feel books are for using, for learning and enjoying, not just looking pretty on the shelf, like so many boat anchors. Besides shelf space is limited. Books I no longer need or care for, I get rid of, keeping when practical, those that might be of use in future. Though you did not ask my book buying habits, I rarely buy so called "top ten best sellers", at a book or grocery store, unless their subject is of extreme interest, if you find any of them on my book shelf, I probably bought them as "remainder books" at a reduced bargain price. A very few books I have, were special ordered, their subject, not being a "hot" seller, or I wanted a certain "edition". Before we get into the main subject of this book I would mention the following, along with man made ugliness, there is a beauty and wonder in the natural world, a child that finds it will never forget it. Life also usually offers unpredictable experiences.

Fascinating Loose Ends and Curiosities

Over the years, some experiences make a special "impression" on a chap, so much so that they stick in the mind, like a sliver, that reminds a person it is still there. Our family doctor, up in years at the time, related such an incident that stuck in his mind, long after it occurred. As a young physician beginning his practice, a patient came in requesting, to please remove a steel phonograph needle he had accidentally sat on, such needles were roughly a tad longer than 1/2 inch, 1.27cm., and a roughly two millimeters in diameters. The young doctor set about the task diligently probing for the needle, but could not locate it, he then sought the advise of an older doctor in the same offices "What should I do? I can't find it?". It was not the fact the patient might not be able to play his favorite 78 revolutions per minute recording of **Beethoven's 5th symphony**, it was the fact that after all that medical school, he still did not know what to do.

It is now seventy years after I myself had an experience that awakened curiosity in my mind. I was about five or six at the time (roughly 1942), my grandfather had bought two live chickens from a farmer for the family dinner. Well first they needed killing so he chopped off their heads in the ten foot square laundry room. He had just left the room when I walked in and stood there amazed, as I watched the headless chickens walking and running around the room for what seemed like five minutes or more, I wondered at the time, **what sort of control was keeping the headless chickens upright?** after all their brain was now somewhere else. (eventually they did topple over and were unable to right themselves), **and what sort of intelligence let them walk and run coordinated steps without a head?** Some look down on chicken intelligence, consider though that chickens have a vocabulary of twenty calls for different purposes.

Note: If you are ever thinking of killing your own chickens for dinner, get ready after the deed is done, to do the messy job of plucking out their feathers, there are roughly over two thousand feathers per chicken.

The second incident occurred some years latter when I was about 12 or 13, in the eighth grade (my birthday, feel mid term). The grade school I attended was associated with a college that taught teachers, so they would have some actual, on the job training, they were given some time assisting an older experienced teacher in a working classroom setting. This practice imparted valuable experience, and let the learner see, if real teaching was "for them". On this particular occasion the new assistant teacher was a young women with a pleasant personality who to put it mildly, was a "knock out", a singularly beautiful woman, she could have been a professional model or movie star, she had a lovely build, red hair, and **canary yellow eyes**. Yes! Yellow eyes, this was in 1950-51 (8th grade). **The question is: Are there people with naturally yellow colored eyes?** In my admittedly limited experience over the years I have never seen another person with yellow eyes, and I venture neither have many optometrists, and no the color was not due to the illness yellow jaundice. Were colored contact lenses even available in 1950? Mass appeal of contact lens use was not achieved until the 1960's, as late as 2004 only three percent of contact lenses purchased were used to change eye color.

The third incident I am reluctant to even mention, and have left it lie quietly all these years like a sleeping dog, never speaking of it even to my father, the reason for my silence being, some persons are all to anxious to label someone that would relate such an experience, a kook, a crank, or an attention seeker. Still it is an interesting curiosity and best of all it is true.

I was between seven and eight years old when this particular incident occurred, it was during WWII in 1944 or 45, I was living in an older part of the city at the time, in a walk-up third floor flat, My grandmother had a lovely two and a half by five foot bed of "Moss Roses" (portulaca grandiflora) in assorted colors out front, bees actually lived in holes in the earth between them.

FIG._1_ Moss Roses (Portulaca Grandiflora) perennial. Height aprox. 5" (12.7cm.)

On this particular day I was kneeling in the small unfinished front attic store room of the flat, an open attic door was to my left, there was a single 40 watt light bulb on the wall next to the door to help brighten the room, to my right on the opposite far wall nine feet away (2.74m.) daylight was streaming through a small double hung window. Overhead the ceiling sloped down toward the outer walls. **I was scratch building a little wooden five inch long (12.7cm.) race car of my own devising** out of pine wood, my work area was on the floor before me, to my right for three or four feet (0.9-1.2m.) lay a clear space, cardboard boxes of stored things were stacked here and there in the remainder of the small attic.

I had been concentrating on the little car in my hands trying to figure a way to fit a curved plastic windshield into it, when nearby on my right I was surprised by a rustling sound, like someone stepping on dry crumpled paper. Still looking downward and immediately turning my head toward the apparent source of the distraction my vision swept across the floor toward the sound's origin, It was about three or four feet (1.2m.) away, at the far edge a clear area, there my gaze fell upon **a pair of small feet** pointing toward me, as my gaze began to rise from the feet, legs up to the knees came into view and a vague unfocused impression of a lower body, grayish in color. I believe the being wore some sort of smooth body clinging costume. I quit raising my gaze any higher because, a distinct and overwhelming idea flooded into my mind, it was, that if I looked higher

A primal fear gripped me as I gazed upon a pair of lower legs and feet.

and into the eyes of the small figure I would be in it's power. This was not a "maybe", it was as certain as the coming of tomorrow. A wave of heat swept through my body, adrenalin had raised my body temperature in the time it takes to count to two. In response I slowly stood upright keeping my eyes looking downward, turned away from the small figure and proceeded down the fifteen foot (4.6m.) long hall toward the kitchen where my mother was preparing supper. She was standing by the sink as afternoon sunlight fell across the kitchen table,
I said to her **"Mom, there is something in the attic"** ("something in the attic", was that an under statement.) **"please come and see".** I was rather surprised at the time, that she did not get excited and rush to have a look, maybe mothers get "conditioned", by all the stuff they have to endure raising kids. The two of us then went and looked in the attic, there sat my little race car and the boxes of stored things, but no small person. Twenty years later, I still had a fear of going alone back into that

attic, even though the flat was empty at that time. In no way did I wish to re-live the incident. Some things are hard to shake off.

What did I really see as I knelt in that attic?

All I can say is the room was three stories above street level, had only one door, and I was directly in front of it, There were no children in the flat other than myself and we had no pets. The previous ten or fifteen minutes before the incident I saw nor heard no one in the room with me.
I was shocked at the time to find a 23 inches (58.4 cm estimated total height) grayish figure standing four feet (1.2m.) away from me. A squirrel standing on it's hind legs would have been to short. As for the little race car I never did get the curved windshield installed or the CO2 cartridge that was to power it like a rocket on wheels, Still it rolled around nice and looked good.

The world can by turns, be a bit scary, and a bit strange, but then days come when it is just delightful to be alive, perhaps that is what keeps life interesting. Be this as it may, that "awakened curiosity", about our world has never left me.

The Theory in this book

The theory that unfolds in this and my two previous books **"The Man Who Saw Space and Time"**, and **"I Am Universe", is a theory about there being "dualities", within a singularity**, the singularity of the theory is, **the Cosmic Ocean, the universe**. A basic duality within the cosmic ocean , is **Substance** the ocean's body, and **Energy**, the animator of the ocean's substance. A complementary opposite pair is one in which, the one of the pair, "complements", the other, each gives the other value. Example: "White", has no value without "black", each needs the existence of the other.

Space and Time are complementary opposites, they both are considered a "Substance" in this theory. Space has two halves, one is **the stationary aspect, the other half a moveable aspect, they both occupy the same volume of space simultaneously.** Time has a similar arrangement. The two, space and time are the totality of Substance in our universe, they are like the front and back of a single coin. When energy acts on primordial substance, it's, behavior, and characteristics are revealed.
When the movable half is moved in relation to the non-moveable half it does so by a rotary motion, a torsion motion. The amount in degrees of relative rotary motion between the moveable half and the unmovable half, is termed "displacement". This movement of the halves of Space, or of Time, is propelled by energy when it occurs.

There is a tension between the halves of primordial substance, when displaced it is a longing to be whole again, un-displaced, aligned, as they once were, "to be", once again, a complete Self. For now let us term the theory, "The Matrix Theory". Just as I have been referred to by different names over the years, others will ultimately give this theory a favorite name.

Primordial Substance, and Primordial Energy are together the animated, living body of reality.
Some personify this pair within the whole, as a Goddess, a queen, and a God, a king, and the Cosmic Ocean their kingdom. This is true as far as it goes, however modern psychologists have found individual persons can have a dual behavior, a primary behavior based on their physical gender, and a secondary behavior applied when needed, which is gender wise the complementary opposite of their primary behavior. I believe the cosmic ocean's duality behaves likewise. **Primordial substance** has two characters, or aspects, **Space and Time**, scientists have discovered primordial energy, possesses two characteristics or aspects also, Potential energy, and Kinetic energy. We see the substance of the Cosmic Ocean, by it's **Image,** it's form. The cosmic ocean's **Likeness**, it's power, animates the substance, and gives it a directedness. Energy has been termed, "The Power", "The Will" and "The Spirit". The biblical old testament account commented "The Spirit", of God moved over the waters". Genesis 1, vs 2. The waters are the substance of the cosmic ocean, the universe.

Realizing the value of this theory

The matrix diagrams show the relationship if their components.
"Value", judgments can be made by how close, or remote something is, compared to the extreme aspects of it's type.
Example: On seeing pure black and pure white a person can judge how grey a thing looks. Secondly when the definition of an extreme is made, one

aspect is often defined in terms of what the other is not. Example: Light is the opposite of Black, or Black is the opposite of Light.

With that revelation it is realized that, **Each extreme of a pair of complementary opposites, gives value to the other extreme,** as each helps define the other, because of this each extreme is of equal value, and of equal importance.

Myths, or religions based on such myths, in which the male gender wields all power over the women or the other extreme where women wield all power over men, are not balanced, not individuated. The Shadow Self must have it's due, in the mind, in the legal courts, and in the bedroom.

To achieve balance the "**Action Self**", the "what I say goes" Self (the masculine side), needs to moderate it's activity when conditions warrant, needs to consider what the "**The Re-Action Self**" (the feminine side) believes would be advantageous for both selves.

Part One

The Eternal Cosmic Ocean

Introduction to the Cosmic Ocean

No matter what peoples, or civilizations, expresses themselves, eventually they speak of the same basic things, they give those things various names, and use varying descriptions of them, some see them in terms powers of the natural world, Rain (water), Fire (power), Wind (air), and Earth, the substance beneath their feet. Others see these things in terms of the spirit powers, or personify them as Goddesses and Gods.
Naturalists speak of **"Nature"**. Scientists speak of **"Facts"**. Psychologists speak in terms of **"Behavior"**, Gamblers speak of the odds, and **"Luck".** Mythologists, Priests, Rabbis, Mullahs, and Seers, hedge their bets and use a mixed terminology, they speak of **"The Seen and the Unseen",** a world we see, and an unseen world of spirits, souls, transcendence, and higher and lower natures. **We know, they all speak of the same things because that, is all there is.**

The world may appear complicated, but many believe it's complexity has sprung from an original or underlying simplicity, that our reality evolved from a

few basic components, permanent things out of which, the ever changing world we experience grew.
I believe people have an inner need to find that simplicity and permanence, a need to find a foundation that can be relied on to be there, to find something that will not die and pass out of existence. They long for stability in their lives, a respite however brief in a turbulent world, a world in which even rocks wear away and crumble. This is not a desire for a world without change. Needed change is welcome, like food when we are hungry, or some joy rather than all sorrow.
Man, after generations of contemplation and experience, has found that longed for **"Simplicity"** and the things of **"Permanence",** they were there before him all the while, man had but to recognize them for what they were.
The answer to the question "What is Permanent in our world?", Is **The Cosmic Ocean,** with it's simplicity of **three basic components**.

We probably will never know if the "Cosmic Ocean" has limits to it's volume, we can see it reaches beyond the farthest stars found. What matters is that we understand enough about it's nature to use it to our advantage. I believe man will understand enough one day, to be able to do things which if done today, would seem magical.
"Why does change occur at all?" or put in other words "What animates our world?", and "What regulates", the animation?

All things that are <u>born</u> from the permanent Substance of the Cosmic Ocean, have a limited life span, eventually they die. Plants, Animals,

Viruses, atoms, electrons, yes, even an electron may perish, as when an electron meets a positron and converts to pure radiation (traveling waves instead of resonant spherical, standing waves). One would think that a day might come when all activity, all life would "wear out" and be no more, but that is not so, a mysterious animating and regulating power "regenerates", things, causes them to be "born" a new, what animates nature? Let us see what the thinkers over millennia have arrived at to answer these questions.

Where is the place where Light dwells, and where is the place of Darkness? It is

The Eternal "Cosmic Ocean"

Ancient story tellers explained there exists a vast "Cosmic Ocean", not of the water of our earthly

oceans, but **it's "waters", were of a "Primordial Substance"**. A "Substance", that could be formed into earthly seas, into the atmosphere of our world, the earth beneath our feet. or even formed into the fire of the stars. Now when you look into the night sky on clear nights you will be seeing "The Cosmic Ocean", vast and forbidding, yet reassuring, dependable, eternally there. In fact you are standing in part of the cosmic ocean and are made from it.
Ancient Sumerians, Babylonians, Chaldeans, Egyptians, the Aryans in India and more peoples, concluded that:

> **All things come from and return to this "Cosmic Ocean", an ocean infused with a mysterious all pervasive animating power, and a messenger between the two,** between substance and non-substance (energy).

The Cosmic Ocean is the permanent, that lies behind the impermanent, it is the eternal substratum, of reality. The Cosmic Ocean ebbs and flows with change. The animating power molds substance into "forms", then releases them, returns them, into dissolution to be reformed in the future.

I have speculated as others have that the cosmic ocean is all that exists, it is a self contained reality, outside of the cosmic ocean there is no Time "To Be", no Space to exist in, outside of universe there is no animating power (Energy), to bring change. The Cosmic Ocean, is a singularity, a unity, a whole. As a Judge in British author John Mortimer's fictional tales of "Rumpole of the Bailey", might have termed this situation, The cosmic ocean is Sui generic. Sui generic is Latin, and means, unique, in a class by itself.

The waters of the eternal Cosmic Ocean, are commented on in various sources. The biblical book of Genesis, simply terms them, "**The Waters".** They are the formless darkness of The Rig Veda.

"In the beginning darkness existed enveloped in darkness. All this was indistinguishable water". Rig Veda, Mandala, X 129.

This was a formless cosmic ocean, in motion with no rigidity, a void.

The Laws of Manu state: **"The Void"** was "indefinable, as if immersed in sleep".
In **The Voluspa**, we find it was time's morning when **Ymer **** lived, there was no sand, no sea, no earth, no lofty heaven, only the **Gulf of Ginnuga,** the yawning gulf, or ginunga gap. The Gulf of Ginnuga, is the place where the cosmic ocean's waters are dark and formlessness. The Voluspa is also called Sybil's prophecy, an Icelandic tale of the history of the world of gods, men, and monsters, from the beginning to the twilight of the gods (The Ragnarok). At this, twilight of the gods, even the stars will vanish.
The giant **Ymer** (or Ymir) of the Voluspa tale, was a primal being, the first being, Ymer was not a man. Ymer not being a man brings to mind the image of a Golem, In Jewish folklore, **a Golem** is a lifeless clay image brought to life by a spell. In Genesis, Chapter 2, we find man himself was formed out of clay, and

made a living being by the creative power. Did the wizard God in the Genesis story make Golems? The Genesis creation account is another story attempting to explain how living creatures were made from seemingly inanimate substance.

The Cosmic Ocean is still a very good basis to understanding the reality we find ourselves in.

"The waters" of a formless ancient "Cosmic Ocean", are described as in darkness, a **Darkness of Chaos.** The darkness spoken about was not duc to a lack of sunlight. "The Light" of the biblical Genesis account on the first day of creation, was not the "light from celestial bodies, the stars, the sun, or moon. Celestial bodies do not appear in this Genesis story, until the fourth day.

The Light spoken was, **a Cosmic Light, a power**, that can reveal Order (**Cosmos**), from out of Disorder (**Chaos**). The restless animating power that inhabits the cosmic ocean was yet to separate "order" from the "disorder", of formlessness. in this mythical beginning, there were no "forms" to be discerned by the mind of man, only randomness, thus the mind was, as in the darkness of unknowing, it could not experience, that is see, what was not there.

Random activity of substance is driven by the mysterious animating power, eventually "forms" resulted, and the chaos, became less chaotic. A prophet might say, The cosmic light drove out the darkness of disorder. Much as the mind's power to reason organizes random experiences, random data, to reveal an order.

Duality in the midst of unity

Duality and the Complementary opposites

A Duality in this book, is **a pair of complementary opposites**, one being what the other is not. The most basic duality is where the cosmic ocean "is", and where it "is not". Outside of the cosmic ocean is where it is not, outside of it is nothingness, a lack of existence. Within the Cosmic Ocean there are many dualities. **Within the Cosmic Ocean,** is the duality that is it's substratum **"Primordial Substance"** and a **"Non-Substance".** The non-substance is Primordial Energy**, an Ability**, to animate substance
The cosmic ocean can be pictured as a huge sphere surrounded by nothingness. **All substance and power is within this sphere,** outside of it, there is no space, no time, no energy, no-thing. A sphere with nothing beyond it. You may hear a little nagging voice within you saying, "but, but, what if I went beyond, outside of the cosmic ocean sphere?" There is no "there to be in" outside the cosmic ocean sphere, you and the "outside", would be nothingness, in nothingness, nonexistence.

The author of Biblical Genesis put the duality of the cosmic ocean this way: "...darkness was on the face of the deep, and the spirit of God moved over "the waters". Ch. 1 vs. 2.
In other words, before creation **a substance "The Waters"** existed, and also, a Spirit, a **non-substance,** an ability, a power, existed.
That is the view the theory of this book takes.

Note: In my writings, the Cosmic Ocean, the "Universe", the Void, and Brahman, are the same reality, "The one and only Reality", "The Totality of Existence".

The Cycle of existence of gods and men.

From Teutonic mythology we are told, at the time of **the twilight of the Gods, the Gotterdammerung,** our world, became a world in chaos. After the battle, the Gods regained control of the forces of nature that gave the world shape and order, the waters of a great flood receded, green meadows appeared, and the Eagle again was seen flying and hovering in the sky.

The whole of the cosmic ocean appears to be engaged in a similar eternally repeating cycle of change. A cycle of change from **Chaos to Cosmos and then Cosmos to Chaos**, is a sequence of increasing and decreasing percentages of each state it is a "**Universe Cycle",** one extreme, chaos being disorderly and cosmos being it's complementary opposite orderly.
In a complementary opposite pair, **what one is, the other is not**, they are extreme states or conditions of existence, as are the Chinese "qualities", Yin and Yang. Both exist simultaneously, only the percentages of each change over time.

Chaos - disorder, randomness, aimlessness, a jumble. A state in which "chance" is the dominating controller, a state without certainty, the appearance of form and it's duration is unpredictable. Chaos has the **"makings"**, the potential to become all things, the ingredients but no predictability over time of becoming them. Chaos is the complementary opposite of Cosmos. **Chaos is infinite diversity within it's whole. Infinite Complexity, a jumble of existence**

Cosmos - order, harmony. The Cosmic Ocean when in a harmonious orderly state, a state that has sequence, and predictable operation, where certainty controls and encourages order. It is a world of repeated forms, and matched pairs of complementary opposites. **Cosmos (order), is infinite simplicity as a whole, a unity, a oneness.**

Note: - examples of different matrix diagrams in this book are titled: "The Cosmic Ocean" (Universe), "The Self", and "The Mind".

Other Examples of Duality in the midst of unity. pairs of Complementary Opposites

In a pair of Complementary Opposites each of the pair gives "value", to the other, each gives the other reason for being, each is a reaction to the other's action, each is, what the other is not. Together these extremes of their kind, form a whole, a completeness of their kind.
The most basic pair of Complementary Opposites in the Cosmic Ocean (the Universe) is **"Substance and Non-Substance".** This is the first pair on the list of complementary opposite examples to follow.

a) **Substance and Non-substance** (Energy)
b) **Actuality and Potentiality**

c) **Chaos** (Disorder) **and Order** (Cosmos)
d) **Change and Permanence**
e) **Actuality and Potentiality**
f) **Action and Re-Action**
g) **Male and Female**
h) **Satisfaction and Need**
i) **Self and Other,** the twin persons, the one we believe we are, and one we believe we are not.
j) **The Ideal, and the Not-Ideal,** the twin lands, one a land of "potentiality", the other a land of "actuality".

They exist in the mind. Our mind looks for, and looks at these pairs, arranges them, compares them, to make sense of our world, to see how things are related to one another. Many other complementary opposite pairs exist here are a few:

Light and Dark
Known, and Unknown, (the seen, and the unseen)
To have and To have not
Same and Different
Active and Passive
One and Many
The quick (living) and the Dead (non-living)
Full and Empty
Whole (together), and Separate (apart).
Push and Pull
Yes and No
True and False
Odd and Even
High and Low
On and Off

Gender within the Cosmic Ocean

Various myths have assigned the cosmic ocean, characteristics of sexuality, a gender, to explain how a single being, entity, or thing, can give birth to many things.
A myth of the cosmic ocean that sees it as containing within itself both male and female genders explains such a thing. A hermaphrodite universe. It makes sense but that does not mean it has both a big vagina and big penis. To my mind picturing that, is rather an unnatural perversion of the real situation. A better way to see this sort of situation is as follows.

The substance of the cosmic ocean, can be seen as, "The mother of all things", Substance in it's entirety is **"The Womb of Creation"**, in other words as feminine. One could visualize this as a feminine being, personified as "**The Queen of Heaven**".
"**The Power**", the energy, that brings substance to life, that animates the cosmic ocean' is a non-substance, an ability once termed "The Power", today scientists term it energy. This all pervasive power "the non-substance has been visualized as masculine, personified it can be visualized as "**a king ruling his kingdom**". Together they comprise a heavenly couple. Eventually they conceive and give birth to all things.

The child of this cosmic queen and king, is their messenger, this child's gender is a reflection of it's parents, it can have the strength of it's father and the gentleness of it's mother. The child's strength is "force" and it's gentleness, is counterforce. **These**

interpretations of gender, within the cosmic ocean are those of the theory in this book. Also see "Goddesses and Gods", and "Gender", in the Glossary.

Psychologists have explained behavior in humans as being of two aspects. 1) A primary behavior befitting their physical gender and the expectations of society, 2) a complementary opposite secondary behavior used in reaction to urgent need during an encountered situation, a "shadow behavior".
I propose the three basic components of the cosmic ocean, each have two aspects to their behavior, a primary and a secondary aspect.

I believe Both Primordial Substance and Primordial Energy, like Man have a Shadow Self, which they exhibit as needed to bring about dynamic balance. Bring about a kind of dynamic harmony as ideally exists between men and women, or between the halves of an electromagnetic wave.
Primordial Substance is feminine, Primordial Energy is masculine, these two primary genders are their physical sexuality and primary behavior. As with humans each gender has two aspects to it's behavior, one aspect behaves as their physical gender is expected to, the other aspect, is a shadow behavior of opposite gender.
Substance as Time, and a**ctualized** energy (kinetic energy) are of masculine aspect in behavior.
Substance as Space, and **Potential** Energy (held energy), are of feminine aspect in behavior.

Existence with Limitations

A Cycle of existence

As Teutonic myth noted, even gods could perish and be reborn.

The Cosmic Ocean is Eternal, but the things made from within, and by it are not eternal, one day they dissolve back into the Ocean from which they were born. Each "born", thing, has a "life-span", a duration.

A Cycle has a beginning and an end. Some picture the cycle as having four stages. A cycle of life, has been described as having a, P**lanning stage**, potential goals (Atma), **Selecting a specific goal.** (Buddhi). An **Implementation stage** (Kama), the actualizing of change, on-going efforts to bring about the desired result, and Sthula, the behavior, (characteristics) of **the formed substance as it is made a moment at a time.** This last, is stage, is the accomplishing of a goal over time, that was visualized in the Atma stage. the goal is achieved by a cycle of change, the goal's potentiality in Atma, becomes actuality, in Sthula, a moment at a time.

All that is born, comes from, and returns back to, The Cosmic Ocean. The Sea of Being.

FIG.__2__ I looked and saw amongst the burning Stars, this Island Earth, this blue planet, floating in the Void, floating in, The Cosmic Ocean.

The Nature of the Cosmic Ocean

The cosmic ocean is a whole, a **singularity**, within which are intimately associated **dualities**, dualities of extremes with tensions between them, yearnings between them, strivings between them, striving to one day be together, and yet each partner of the duality wants to exist as their own "self". The dualities as a group, operate like **a "gestalt", which is, many minds working as one, each linked to and affecting the others.** Each duality's tension, is linked to all the others. Looked at another way, the dualities form **an entanglement, they experience one another's presence and condition**.

The nature of the cosmic ocean is one of differences, extremes that are, complementary opposites. These dualities are imbalances, one is what the other is not. Imbalance, drives change, imbalance strives to be balanced, strives for a "leveling out, an equalizing". I say **"strives"** because as long as dualities exist imbalance exists. **More** change here, and **less** there, **faster** change here, and **slower** change there, **a direction** of change here, and the opposite direction there, resulting in a striving to attain **the ultimate desire of a "dynamic equality", of existence**.
The extremes "Strive", for a goal of **"Dynamic Balance", "Dynamic Equality"**, a goal that is always just beyond attaining, because the extremes of the duality sway back and forth between two goals. **Each extreme of the duality longs to be unified** with it's opposite, **yet each extreme, also longs to be an individual**. It is that way with individual human genders, and the same with all dualities in the cosmic ocean.

Dynamic Equality it can only be achieved by the extremes of a duality by sharing existence, a teeter-tottering, an alternation, where one extreme becomes more for a while, and the other less, then reversing, so the more becomes the lesser, and the lesser becomes the more. **See note below**.
This striving by alternation is continually repeated, to maintain their mutually shared existence, and yet be a duality.

Like the existence of **Yin**, and **Yang**. One extreme has little value without the other. Each yearns for a measure of separateness, to be individual, and when separate they each yearns for a measure of being together, a measure of oneness, a measure of being whole.

The cosmic ocean as a whole, is locked in an on-going cycle of striving for equality, swinging like a pendulum from a condition of chaos (disorder), to a condition of cosmos (order), then cosmos to chaos, eternally repeating.

Note: In the mechanics of vibrating systems or electronic circuits when trying to reach a desired condition, there is a tendency to repeatedly "Overshoot" the mark, then "Under shoot", the "Mark", the mark being the desired condition. This process is termed "Hunting", This "Hunting", situation is generally undesirable as it can lead to destructive vibrations and system failure.
Though "Dynamic Balance", is always being sought, always "striven", for, there is a useful compromise sometimes attained, a non-destructive condition, wherein each extreme has a life-span, a cycle of modest dominance, then hands over modest dominance to the other extreme, this non destructive stable condition termed **"Resonance"**. Musical instruments and mechanical systems can "Resonate", with pleasing results. The electron, is a resonating wave structure, it has a very long life span and good stability.

The Three Basic Components of The Cosmic Ocean

The three components are **"Substance", "Energy", and** a mysterious third component, an agency, that allows the interaction of the first two components, let us term it **"The Connectedness"**. The "Substance", of the Cosmic Ocean consists of "**Space** Substance", and "**Time** Substance".

Space and Time the two "aspects", both aspects together are the whole of primordial substance.
Energy, is a non-substance, the complementary opposite of primordial substance, it is an ability. Energy has two aspects, one kinetic, the other potential.
Primordial substance has a **"Connectedness"**, a. continuousness, an extension without void or break. This connectedness, works like a spider's web, it transmits, or carries forces, and the message

of force, throughout substance, throughout the universe. The "connectedness", lets the aspects of energy, affect substance, flow within substance from place to place.

The third Component is, an agency, a facilitator of the interaction of the complementary opposite pair, substance and non-substance, it allows energy to be exchanged, allows energy's efforts to move within substance.

All born things and all behavior are due to the interaction of these three basic components. These three components have three natural inclinations, or "needs":

1) **to be whole.** 2) **To be individual,** and 3) **To share existence.**

Substance the first component of the Cosmic Ocean.

Illusive Substance
Space and Time

The underlying substance from which all things are born, is primordial substance, it is the body of the cosmic ocean. Primordial Substance has two aspects, space, and time. Substance fills the Cosmic Ocean, and is this ocean's only substance, it is homogeneous, grain-less and continuous (touches itself throughout), it is without voids, cannot be made more dense, compressed, or less dense (rarefied), yet is elastic, and can be displaced by "stress forces". When "stress forces", are removed, the "strain forces", within displaced substance subside as substance returns to it's un-displaced condition.

The Primordial Substance is not a gas, vapor or fog, and will not change state at any pressure or temperature. With the application of Energy (a non substance), the behavior of primordial substance changes, some substance displaced substance by energy becomes the chemical elements, some becomes "fields of force". **The two aspects of substance are "space",** and **"time"**. Space is extension and potential form. Time is an actuality as, a duration of existence, within which change can take place. Energy drives change, drives motion. These dual aspects of the cosmic ocean, **substance and energy** are both real. The potential of space is "real" as a potentiality. Time is an "actuality", and is real as an actuality".

In the theory in this book

Primordial Substance has the characteristic of being **"Displaceable"** (as described in the text and glossary). Displaced, Primordial Substance is "Strained" Primordial Substance. This condition of strain can be interpreted as the **"forms"** we experience, as particles of matter, as wave bundles, or as the chemical elements. Un-displaced substance is formless.

> **Primordial Substance, is half of the duality "Substance-Energy".**
> **Primordial Substance is the only substance that exists.**
> **Primordial Substance, has two aspects or states, Space and Time.**
> **Space and Time**, are complementary opposites in behavior, one is, what the other is not, and contra wise. Space has potentiality and can be "molded", into all born "things".
> Time is an on-going actuality, it is formless, and lets change flow.
> **Space,** is primordial substance with potentiality
> **Time,** is primordial substance as an on-going actuality.

The following about primordial substance appears to be so, it **is eternal, is not "made", is indestructible", and for all I know, is infinite in extent.** The cosmic ocean is one vast eternal singularity, a volume of substance infused with energy.
Let us now look at the two aspects or states of primordial substance, Space and Time.

Space

Eternal Space, an aspect of primordial substance

Here in this book, Space is a characteristic, or aspect, of eternal substance. **Space is place, and when "molded" by energy, space takes on "form"** (shape).

Space appears to be a substance with elasticity.

Space has usefully been resolved into three dimensions, three directions of extension, these describe an object's form, relative position and relative size. The names of the three dimensions of space, are **Height**, **Width**, and **Depth**, (each at a ninety degree right angle from the others).

The following has been said of space:

Elastic Space
Space moldable by energy

Space and Time, are Primordial Substance, they are the "physical stuff", of the Cosmic Ocean, it's body. The complementary opposite of Substance is **The Non-Substance,** Primordial Energy**.** Potential and Kinetic Energy are energy's two aspects. In the theory in this book, Space, consists of two halves, **one half is not moveable**, not compressible, expandable (rarefiable), or separable into particles. Space itself remains continuous, unbroken into pieces. What appear to be particles is the effect of the aspects of energy displacing space, causing wave activity. **The other half of Space** is moveable, is rotate able, as a torsion motion. When the movable half is twisted (by rotational motion) in relation to the immoveable half, it is said to be **"displaced"**.

When space is displaced a condition of "strain", exists within it. A tension between the displacing effort force, **the "stress force"**, and the "reaction force", **the "strain force"**. The strain within, caused by displacement of Substance tends to return the moveable half of space back into alignment with the non-moveable half when the displacing force (stressing force) is removed or reduced.

What we understand as "**form**", or "**shape**" is due to differing locations and levels of strains in space substance. Matter is a special condition of strained space, a standing wave structure. Matter has the property of "inertia", inertia being the ability to resist a change in it's bulk motion or position. Primordial Substance that is not matter, only has resistance to having it's displacement increased.

In the field of mechanics, when an elastic substance is bent or forged, it becomes strained, the **strains hold potential energy**, It is strain in the material of a spring that lets it "spring back", after it has been bent, twisted, compressed, or stretched.

When space is not displaced. When no strain is present and no fields of force are present, space is formless.

About "Limiting Velocity"

Though Space in this theory is "displaceable" it is not compressible or rarefiable, that is space cannot be made more dense in some places and less dense in others. Both Space and Time Substance each totally fill the universe. The reason light waves seem to have a limiting velocity, is that, the displacement of the space substance varies little across the vast distances of interstellar space. This means that the strain due to displacement, changes very little over those distances and takes eons to change it's amount of displacement. **The actual velocity of waves** propagating through so-called empty space **is due I believe, to the level of "strain", in that space.**

The level if strain in space, surrounding celestial bodies gravitational fields, would according to the theory in this book, affect the velocity of electromagnetic waves passing through them. It probably would curve the path of light waves passing through a spherical field, due to variation of gravitational field strengths at different distances from the body. The situation is quite different within a resonating wave structure, like that of, an electron or atom, where the level of "displacement", and velocity of displacement **within it's volume**, varies exceedingly rapidly.

Questions about space.

a) - In what Space, do the Goddesses and the Gods Live?
Answer: In the mind of the thinker. In the space of one's thoughts.

b) - Is Space more than an empty container with extension?
Answer: Yes. I believe Space has characteristics of behavior and can take-on "Form".

c) - Does space have "elasticity",?
Answer: I believe Space has elasticity, and can be distorted with a torsion motion, by energy.

d) - Is the volume of the Cosmic Ocean limited in extent?
Answer: never having been to it's limits, I can only ask, "What if" it does, or does not have a limit?

Below are some of the things people have said about "Space".

Space
Real Space
Spaced Out
Outer Space
Finding Space
No Space, No Shape
Like Air I need Space
"Down into that Space"
"Where did all this Space come from?"

"Can this Space volume, be moved over there?"
or put in other words: "when we move a solid object does the space within it push aside the space it is moving through?"

"Space is the stature of God."
Joseph Joubert

Time
The Hands of Time touch Tomorrow, Today and Yesterday

In this book Time is a characteristic or aspect of primordial substance. Time is the complementary opposite of Space. Time is what space is not. Time is formless, and un-formable. There are those who believe time is a fiction, and the future does not exist as an actuality, that the future is always a

Potentiality. These persons believe that only "The Now", is an Actuality.

I see time as a reality of existence.

I see time as a characteristic of Primordial Substance

Time, like Space can be "displaced"

Picture time as one side of a coin, and space as the other side, displacing one displaces the other, both sides turn together.

Movement through Space brings movement through Time.

Time is a duration, an extended moment, "Duration", is a series of "Nows", or it can be seen as a personal "now", moving along the band of eternity, **during which, an action, a condition, or process, can take place. Time is eternal, it always was, and always will be.**

"Distance", or "Extension", is a moving point, or it can be seen as, "the substance of one's being", amidst the vastness of the cosmic ocean.

Some say, Time has only one dimension, and they have named it, the fourth dimension, according to this view, the other three dimensions are those of space. However, a dimension being a measure of something, Time has three useful dimensions all it's own, useful measures to show relationship of events or experiences, The Past, The Present, and The Future, all can be measured, and compared to one another. Time's dimensions are unique however, for the can show, a **"location"**, in a **"sequence"**, of events **and** their **"duration"**. Whereas Space's dimensions are unique in their own way, in that they show the relationships, of relative **"place"**, relative **shape**, and relative **size,** of objects.

The book "Chronos", is a very nice piece of work on the subject of Time. (Thunder's Mouth Press, by Etienne Klein, c 2005, ISBN #1-56025-708-3 & # 13: 978-1- 56025- 708-0, an Imprint of Avalon Publishing Group). In that book author Klein questioned what time was, before the process of actualization? that is, what time was before motion took place. Put another way, when nothing is happening, does time exist?

For my part I believe **Time is always in existence, always "on-going", always "actualized",** that the moment we term, a "now", a moment before "now", or a moment after "now", are actually moments along the band of eternity shared by things simultaneously in existence during those moments. **A certain instant that a person declares as "now"**, has a special meaning to that person, it, relates that person, to other persons, objects, places, or events that occur simultaneously in that special "now". A series of personal "nows", lets a person judge the relative duration of different events, or to allow a person to compare the life spans of different things.

However, **even if no change were taking place I believe "Time", being an eternal Substance, would always exist**. Imagine a coin with two blank faces, one face representing un-displaced, unformed, "Space Substance", The path of a light wave can be bent on encountering a zone of differing "strain", level as it is on passing from air through a glass prism.

Space substance is in a constant state of "potentiality", always has the possibly of it's "strain", level being changed, space has the potential for different behavior, or different "form", as it is molded by energy.

Time facilitates change, facilitates motion of substance, facilitates learning, and facilitates opportunity. Time it has been said, is the best teacher, unfortunately it also facilitates the death of all it's pupils.
The "Now", stays still for no man.

Below are some of the things people have said about "Time".

Time

Time after Time it happens:
Where did all the Time go?
They "Loose Time"
They "Have Time"
The Time is Now
If I had the Time
I'll "Make Time"
Timed Out
Time Flies
Doing Time
The Time Is Right
"Time marches on"
Time can be "gained"
It is like "Time in a bottle"
I have "Time on my hands"
"The Time just got away from me"
"There is no Time like the present"
and they may say "Time is on their side"

On thinking about the nature of Time one can ask:

If we are at "The Now", where is Tomorrow? or for that matter, where is Yesterday? The past? **Yesterday's Gone, Yesterday's Gone**. Yesterday is in the substance of past "nows", the substance of the future "nows", are yet to come.

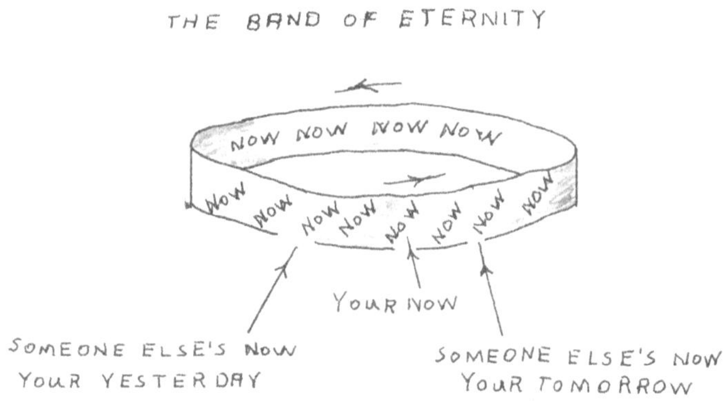

FIG.__3__ The Unending Sequence of "Nows". All the "Nows".

Your yesterday is someone else's "Now", just as your tomorrow is someone else's "Now". We each live the "nows", of our life span.

Time is a characteristic, an aspect *, of eternal substance, just as Space is a characteristic, an aspect*, of eternal substance.
Time, is the substance of change Time, is "The great Equalizer", "The great Leveler". Time lets the weak and strong, the poor and rich, peasants and kings, come to the same end, their inevitable dissolution, their death.
Time is the "elapse", the "duration", for change to take place in. All **created things have a limited life-**

span, a limited span of time "to be", to exist in, a limited duration "to be". Time, lets the cycles of existence "play-out". Time allows Succession, Cycles, Evolution, and Motion, to take place.

* - **Note:** A definition of aspect can be found in the glossary.

Change is not instantaneous, a period of time is needed for the forces of the universe to increase or decrease. Eternal Time is like a road down which the vehicle of the personal "now", of each existing thing moves. The Egyptians at the time of the pharaohs, termed this "moving now", **"The Living Moment"**. This "moving "now" can also be seen as an "extended moment". This process may be similar to the phenomena of "persistence of Vision" in humans.
Time is a "continuum", **a continuousness of substance, time with no gaps,** from one "moment", to the next. The "nows", of time appear to flow like the particles of water in a river.

Elastic Time

In the theory in this book, Time ", is displaceable by energy. Displaced time is **"strained time".** The process of displacing time is similar to that of displacing substance. Displaced Time or Space substance, can be compared to, a spring that is being "wound", and "un-wound", repeatedly. A thing is propelled into existence by an increase of the displacement of time, and proceeds out of existence, by a decrease in the displacement of time.

Undisturbed, un-displaced, un-strained, "time substance", is like a single **"eternal moment"**, a moment that never ends, a time singularity within which all moments of time exist.

Non-Time is the absence of time, to say such a state can exist is a misinterpretation of reality.

Comment on relativity theory:

In the theory I have presented, space is not "curved by gravity" as some proponents of relativity theory speculate. Variation of light wave velocity is due to the light passing through regions of different "Strain", level in the substance of space.
Gravity, is due to reduced strain immediately surrounding resonating "matter waves", such as exist in the chemical elements. Matter in outer space, such as a planet is attracted toward another planet, because each planet is a zone of reduced strain, two planets are pushed or squeezed toward one another, by the higher surrounding strain levels in outer space. It is as if a reduced pressure zone existed between them, a partial vacuum into which they were being sucked.
The standard effect termed "gravity" is viewed as an attraction, it's strength varies with distance from planet surface, strength of attraction reduces as one moves outward from planet surface, by the inverse square relationship. Matter such as the electron is **a standing wave** in which **it's outward face, is a velocity anti-node,** displacement of space in that zone is near zero, and so is almost "unstrained". **At**

the center of an electron standing wave, velocity would vary little and so **it is a velocity node**.
I believe, the lengthening or shortening of matter as it approaches light speed, is due to an internal Doppler effect, in the electron wave structure's shape, as the electron on moving through space, and approaches light speed velocity, an internal Doppler effect occurs to it's spherical standing wave dimensions. The forward portion of the spherical wave, ahead of the wave center is compressed, the trailing portion of the wave behind it's center is elongated. The overall shape of the electron's "standing matter wave", at near light velocity becomes a sort of egg shape.
Note: Space itself is not compressed or expanded only the wave's dimensions in space are.

Space and Time Together

Time and Space are, basic characteristics, or aspects of primordial substance in this theory (see glossary, for specific definition of word "aspect"). Take away time, and eternity is not, take away time and "the now" is not. Take away substance and non-substance and Brahman is not. **"Time", and "Space", are a duality in the midst of a unity,** the unity is the cosmic ocean. They are a continuum, a continuous reality. Time as eternity is an endless succession of moments. Space with infinite extension, is all places, and every place. Without time, the flow of change could not be, and the universe could not live, things could not assume different forms. Change "takes time", no matter how brief.

The Second Basic Component of the Cosmic Ocean. Energy

Energy is, an ability, a non substance. Energy is an ability to bring change, the ability to animate primordial substance, bring substance to life. Energy is the Spirit of the cosmic ocean.
The realm energy, has two aspects, two urges, or states, **the potential, an urge "to unify"**, and **the actually** of an urge to be **"separate"**, between these two "energy extremes", or aspects, lies **"a zone of instability"**, a zone of alternation, favoring one then the other extreme, repeatedly, in this a way "sharing existence".

All change in the cosmic ocean is driven by Energy. Energy is the ability to "displace" primordial Substance,

Substance undergoes displacement by **stress forces,** displaced substance can hold energy, in a field of tension, a field of **strain forces,** within substance.
Stress forces, displace primordial substance. The "Held", strain forces in displaced substance tend to return substance back to an un-displaced condition if stressing forces are reduced or removed.
Energy can travel in waves in the substance of the Cosmic Ocean.

These energy animated waves under certain conditions can achieve a lossless resonance, this resonance activity can be compared to a rubber ball that continues to eternally bounce after being dropped, (however a ball of actual rubber has losses during each bounce, each bounce is a bit less in height than the preceding bounce). During the bouncing process, energy in the ball alternates between kinetic energy of motion, and held energy, potential energy, stored as strains in the ball's deformed elastic rubber. This same energy exchange, kinetic to potential, and potential to kinetic, occurs in the cosmic ocean's substance waves. An example of a resonant wave activity with little loss, is the electron. Does the electron resonate between the space volume within itself and that of surrounding space, the surrounding space acting to reflect the electron wave back on itself?
Or does the elastic space substance of the electron act like the elastic rubber of a rubber ball?

The main features of energy.

a) - **Energy is eternal**, without beginning or end.
b) - **Primordial Energy's** complementary oppo-site is primordial substance. What energy is, substance is not, and contra-wise.
c) - **Energy is a non-substance, an "Ability",** an ability to cause change, an ability to do work.
d) - **Energy does not "wear-out"**, it is a tireless, a restless power.
e) - **Energy is continuous, a continuum,** with no voids or breaks.
f) - **Energy** and substance are never apart, always together.
g) - **Energy is a duality.** The energy of moving substance, and the energy "held" as strain forces in displaced substance. These are the two aspects or "behaviors", of energy.
h) - The two extreme aspects of energy move through substance like clouds in the sky. Clouds of "held", strain forces, and clouds of moving substance that posses energy due to their motion, possess the ability to "stress", form, or displace, other substance clouds.

Energy in moving substance, is termed, **"kinetic energy"**, it is energy which exerts "stress forces", on substance.
Energy is "held", in displaced substance as "strain forces", is termed **"potential energy"**. Strain forces give substance the ability to "spring back", to it's un-displaced condition when the displacing "stress forces", are removed".

The regulation or control of energy, to give energy direction and a goal, is considered separately, and included under the heading **"The Heart of Reality"**.

Once the ability to bring change and the ability to give it direction were mashed, or mixed together under the title **Buddhi**, The Will, and The Logos. The regulation of energy, uses energy to do the regulating, an example of this is the decision making device, the electronic computer, it uses energy to operate.
However in nature, a selection process, can take the place of logic, or thought. The ability "to choose", can be exercised by a confrontation of "attraction and repulsion forces", "counterforce and force", Logic uses "yes", or "no", gates. Prior to "Choice", a comparison of the involved parties is engaged in. A Comparison gives a means to judge, to value, or weigh how far from an extreme a thing is, and determine the polarity of other parties, and judge how stabile a situation is. Electric, or magnetic, polarity, or gender can be used, to supply attractions or repulsions, to sort or orient the involved parties.

Please Note: I here do not say, that energy flows through substance, or that it is "put into, or taken out of" substance. **Energy is always in substance**. Stated in other words, **where substance exists, energy exists**.
That which flows through substance, and within substance, **is energy's aspects**. Potential energy and Kinetic energy flow from one part of substance to another. This shifting and flowing, of energy,

"molds", substance, altering it's behavior, it's character.
The two forms, or aspects, of energy exist. One aspect of energy is **"Kinetic Energy",** this is on-going velocity of substance, in colliding with substance of a higher or lower velocity, it can apply "stress forces", to that substance, The other aspect of energy is, **"Potential Energy",** it is a "held", or stored form, held in elastic strain forces in deformed or displaced substance.
The substance of the cosmic ocean, whether it is understood as particles, wave packets, or fields of energy, when in unpredictable chaotic motion jiggles substance about, one region jostling and bumping adjacent regions of substance. When the animated substance is in harmonious orderly, wave motion, it's two aspects, kinetic and potential, are 'dynamically balanced. E**nergy is always restless, and infused in substance.** Energy does not die, or wear-out, it merely rolls from one aspect into it's other aspect, or it's aspects loose or gain amplitude.

Note: In this book **"Energy", "Will",** and "**Spirit**", are used interchangeably, in that all impart animating energy, or power, that makes substance "to live".
Energy animates all substance, causes substance "to live", in this it is like the wind. Scientists say a thing has energy if it is moving or can impart motion.
Will, is the power of the mind "to think", to make decisions, to judge, to reason, show emotion, and be wise.
Spirit as used in myth, is the power of life which God gives to animals, and man, giving them "the breath of life", when God takes that power back, the

animal or man dies. However in myth "Spirit", also has the ability to show emotions in common with the power of the mind.

Kinetic energy

Substance with velocity, is substance with **Kinetic Energy.** When acting on substance the **"Stress"** forces shape, form, or displace the substance acted upon. Note: see glossary for description of "displacement" of substance.

FIG.__4__ Transferring Kinetic Energy

Potential Energy is "held", in Strain Forces
Displaced substance is "strained" Substance

Reaction force (elastic **"Strain",** forces) within substance, are caused by the displacement or the molding of the substance, distortions in substance. These strain forces are **Potential Energy**, potential in that if released from being held they would revert back into "The Wind of Change", back into kinetic energy, the energy of motion of substance.

Resonance is an oscillation between kinetic and potential energy, an alternation between,

The waves in resonance are half Energeia (kinetic energy), and half **Dynamis** (potential energy). Half of the energy in a wave, can be seen as **on-going change, the velocity of substance**, and the other half of the energy in a wave, can be seen as a **re-action to on-going change, as an inertia effect due to acceleration of substance.**

Stating this another way, **Stress Forces** are the forces driving a spring into compression, and **Strain Forces** are the elastic forces in the material of the spring due to it's compression. Strain Forces are reaction forces, held within substance due to it's having been deformed or displaced.

Potential Energy

Potential Energy Helps Hold The Shape Of The World
FIG._5__

In the theory in this book the waves in primordial substance are transverse waves, these are waves that pulse from side to side, perpendicular to the direction of travel. These transverse waves are in the form of torsion waves, they increase in amplitude of twist and decrease in amplitude of twist, as they advance through space, their direction of twist is alternately in one direction and then in the other repeatedly. They propagate outward, away from their source, unless reflected.

An **electron is a "standing wave"**, a resonating spherical torsion wave, it's clockwise and counter clockwise torsion motions are it's transverse motions, they propagate outward until reflected.

In a standing spherical torsion wave, the center acts as a reflector, **a velocity node**. this is where velocity flow is minimum. The outer circumference acts like the open end of a horn, **a velocity anti-node**, this is where velocity flow is maximum in a standing wave. If there are no losses in this standing wave, it's nodes and anti nodes, would exist forever. To explain again, in other words, in a wave, **"Stress Forces"** are imparted by on-going moving substance, **substance with "velocity".** Moving substance contains kinetic energy. A wave's **"Strain Forces"** are **a reaction to a change in the Displacement of substance, an acceleration.** Displaced substance contains potential energy.

Energy and the Individual

In terms of an individual of either gender, potential energy is possessed by, an individual that has achieved fulfillment, achieved "satisfaction", and become whole, united with that which they were lacking, become whole with their complementary opposite. Potential energy is the bonding energy that maintains the unified state, it is held energy. **Complementary opposite features tend to attract** individuals together, uniting them. **In terms of an individual of either gender, kinetic energy**, is the energy of striving for fulfillment, seeking satisfaction, exerting a directed effort to achieve a goal. **Non complementary opposite features tend to repel** individuals apart, separating them.

The Third Component of
The Cosmic Ocean
The Child of Duality, The Messenger

The Connectedness
Consciousness
The Link
The Soul
"Force"

In the writings in this book seven names are used to describe this third basic component of the Cosmic Ocean, They are all, names for the same ability, the conduction of power (force applied over time), through substance. The Soul, and Consciousness, are traditional names of the same ability, the conduction of information, the conduction of change in substance.

The Connectedness and Force, The Link
The holder of what is, and the bringer of change

In this book: The third component of the cosmic ocean is "Force", Force is eternal, unseen, magical. Forces transport Power within substance, The Fairies of folk lore call upon the magic of "force", to carry out their wills, and like the fairies themselves force, **can hide by standing very still,** appearing to be invisible, (as potential energy can seem to do), **or reveal themselves by the activity they cause** (as kinetic energy can reveal itself in the effects moving substance can bring about).

Wondrous are the forces of the natural world, like angels they can travel swiftly, or like fairies on warm summer eves, dance by moonlight in a fairy ring. I can see the folk lore fairies dancing in my imagination, and actually visit the **"Fairy Ring"** where they have danced.

On some grassy areas there appears a circular ring of grass, greener than the grass outside of it, or within it, **A Fairy Ring**. Though still rather magical, the ring actually results from a fine network of delicate, white, thread like fibers in the soil beneath the ring, these are part of a mushroom's life cycle, the network of threads spreads in a circle, they digest organic material in the soil, and release nitrogenous substances, the center of the circle dies as the food supply runs out, a ring of threads is left. The released nitrogenous substances cause the grass of a lawn to grow greener over the subsoil ring. The mushroom, Marasmius oreades, "The Fairy Mushroom", and others, grow along the periphery of the ring of fine subsoil threads. You and I can see these mushrooms (if you arrive at the right time of year, or see the ring of greener grass above their subsoil threads. These fairy rings, can increase in diameter from year to year as the threads extend outward seeking more organic matter to digest.

The mythologist sees angels, fairies, goddesses and gods, The doctor of medicine and the psychologist see Consciousness, The meta-physicist sees "The Soul", the physicist sees forces, pushes and pulls. These different terms for the third basic component of the cosmic ocean show it is important in different fields of endeavor.

I have had difficulty defining this third component of the cosmic ocean, a difficulty attested to, by the different definitions, and names, others have given to it. Names for this Third Component are: The Child of Duality, The Messenger, The Connectedness, Consciousness, The Link, The Soul, and "Force".
It is, in fact **an entanglement**, a meeting-interacting place, like two or more roads meeting at an intersection, those looking down the roads of the intersection see different scenes depending on which direction they are looking. Viewers in different fields of endeavor have described the scenes they see each in their own way, hence their different views and the differing names they have given what they saw.

This third component is invisible to our eyes, but detectible by it's effects. **It is, a traveler in and between the worlds of existence.** This third basic component of the cosmic ocean, is "The Child of Duality", spreading **Enlightenment,** the experiencing of existence throughout substance. Substance "feels" the effect of the "Will", The Will is Energy's" Intent". The aspects of the forces of nature are: **Force** (a directed effort to bring change), and **Counterforce** (a reaction to efforts bringing change).

In the realm of the "Wind of Change", Force, it's two aspects, or states, are "a re-action to an effort to bring change, a counterforce to any change of velocity (in physics this is termed acceleration). The second aspect is, an effort to bring change, using the on-going Velocity of substance. Between these two extremes of effort, lies a zone of indecision and resolution, a zone of emotion in the mind, and an alternation, which is **a striving for "Dynamic Balance".**

The dual aspects of the cosmic ocean are both real. That which is potential, is "real", as a potentiality. That which is actual is "real", as an actuality".

The third "Basic Component of the Cosmic Ocean", **Force and counterforce move within the continuity of primordial substance.** The integrity, the oneness **"The Connectedness",** of substance, facilitates the conduction of force, the flow of change, and communication within the whole of the cosmic ocean. One or the other aspect of energy is always in contact with substance, never apart from it, substance is never alone. This Integrity, allows all parts of the cosmic ocean to act like a spider's web, a jiggle in one part or one aspect is "felt" by other parts, or aspects. This ability of parts of substance to "feel" other parts, incline one to term it **"The Seer**

of God". The differences in flowing energized substance, can let some portions of substance assume different characteristics or behaviors, and **exert effort forces on one another.**

This messenger that moves within the continuum of substance, and connects the world of **Actuality** (Kama) to the world of **Potentiality** (Atma). Links, the world of **Buddhi** (non-substance, energy, the logos), to the world of **Sthula**, (substance). The continuousness of substance and the forces within it, link what can be", to "what is", Link past "Nows", to the present "Now", and link the moment of the "Now", to "Nows", yet to come.

Communications are variations in the amplitude (strength), of the aspects of energy moving within substance, Some forces are "holders of form", giving shape and character to substance, moving substance is the carrier of directed force. One aspect of force is directed to bring change, the other aspect of force (a counterforce), discourages change.

As concerns **"Consciousness", "The Seer of God", in the mind**, Consciousness is **"The Id"**. The Id, is where on-going urgency is "felt", is "sensed", The two extreme aspects of mind (the Id and the subconscious), are resolved in the "waking conscious", as **"Awareness", and as "Emotion"**.

The Cycle of Change

The interaction of complementary opposite forces, such as "force", and "Counterforce", or "Human "Need", and "Satisfied Need", occurs as a cycle of change, occurs in a flowing wave that possesses the following three characteristics, interpreted in light of the theory in this book, the three characteristics are:

a) **Acceleration**, an increase or decrease in the velocity of a body in motion. Acceleration is here, as **a "Strain force", held within the deformed elastic substance.** and is here looked upon as a **counterforce, a reaction force**.

b) **Velocity** of substance is an actualized on-going effort to bring change.
Moving substance can exert "stress forces" on adjacent substance of different direction and velocity. Stress forces can drive an elastic substance into deformation, or into displacement.

c) **The Displacement of substance.** Displacement is the motion of substance from a reference position to a different position, Stress force brings about the motion of substance.

Substance that has been displaced, holds "Strain forces", within it due to it's being displaced. Stressing Forces can "Form", substance, much as a dropforge can "form", steel. The strain forces within displaced substance can alter it's behavior, it's character.

From a human standpoint **Reaction forces** are the "mores" of society, and personal inhibitions, they are also satisfaction achieved. Re-action forces confront action forces, presenting them with

limitations. These behavioral or character control forces are as real as the push and pull forces of the physical sciences.
From a human standpoint **action forces** are those of human "Needs and Desires", the action forces, strive to achieve satisfaction. Emotion is the resolution of force, and counterforce in the mind.

Genders of the Basic Components of the cosmic ocean

It is quite natural to assume that the substance of the cosmic ocean is feminine as all things are born from it, and that the non-substance, energy, of the cosmic ocean is masculine, the power to animate substance, further, the "link", between substance and energy, can be considered their child, and be of either gender. That "link", is "force", the forces of nature, who's two aspects are, "on-going directed force", and "counter force", a reaction to efforts to bring change.

By considering what psychologists have found concerning human behavior, that humans have a primary physical gender (sexuality), and a secondary behavior of the complementary opposite gender, a shadow self, only used in situations that require it. Considering this I believe each of the three components of the cosmic ocean has two aspects, or states. These pairs of aspects are complementary opposites, as follows:

As concerns Primordial Substance, it's primary aspect is "Space", (feminine gender), and it's secondary aspect is "Time", (masculine gender). Space you will note in the matrix diagram has the potentiality to take-on many forms, whereas Time is an on-going actuality with but one form, with various selected durations.

As concerns Non-substance, energy, it's two aspects are it's primary aspect, Kinetic energy (of masculine gender), the energy possessed by moving substance, and energy's secondary gender, Potential energy, "held" energy", (feminine gender), a reactive aspect.

As concerns the genders of the aspects, force and counter force, the child of substance and energy. Force as on-going directed motion, is velocity, and therefore the primary aspect, and of masculine gender. The secondary aspect of force is, counter force as acceleration, a reaction to the forces of change, and is of feminine gender. The primary and secondary aspects are complementary opposites.
You can view these placed in a matrix diagram Fig.__6__.

The primary aspect or gender is the dominant aspect. The two genders or aspects of each basic component are locked in a cycle that "strives", for "Dynamic Equality", of existence, they will never quite achieve this equality of existence, but continually "strive", for it. Sometime one dominates, and sometimes the other. You might think because these are cyclic processes the results are all very predictable. Not so. The tricky bit comes from the interactions of substance with substance that are random encounters, chance meetings of portions of substance with other portions, as each portion of substance, may contain a different proportion of the two aspects of energy.

Placement of the two "aspects", or genders", of each Basic Component, in a matrix diagram.

In the top horizontal row of FIG.__6__. The row are those things that possess "**Potentiality**", The Substance of Space, is in this top row, as is Potential energy, and Counter-force. Potential Energy, bonds together atoms, molecules, living beings, even to holding together the changing form of a growing embryo in the female womb.

In the bottom row of the matrix diagram are those things which possess "**Actuality**", Time substance, Kinetic energy, and on-going directed Forces to bring change.

I believe **"The Shekhinah",** of myth, is The Goddess, queen of the cosmic ocean, "The Shekhinah", is Primordial Substance, The mother of all things. The king of the cosmic ocean, father of all things is energy, The God, These are complementary opposites, Each gives the other value. Worded differently, each is valueless without the existence of the other.

The child of this queen and King is the messages between them, "Force". Each of these three components has two aspects, one with masculine character, the other with feminine character.

The feminine aspect tends to hold "form". The male aspect tends to be the bringer of form.

You do not need a microscope to find the cause
of many activities in the cosmic ocean.
"Cherchez la femme", "look for the woman".

Please note: In the writings in this book there is a difference between "The Soul", and "The Spirit".

1) - The **Spirit**, is Energy, "The Will", **The Power to animate, and regulate, primordial substance.**

2) - The term **"Soul"** I have reserved for **"the forces", that bring mind and body to life.**

The Soul is the messenger of the power of life, facilitated by "The Connectedness", the "Continuousness", the "Integrity", of primordial substance. Substance is the vehicle that allows the carrying of the forces of nature from their source, throughout primordial substance. The Soul carries energy's messages, messages of changing amplitudes, and changing magnitudes of the two aspects of force. (these aspects are, **1)** the on-going directed forces driving change, and **2)** the "counter-forces", forces trying to "hold" things "as is", so they do not change.

These activities in the soul are **"The Breath of Life", for substance,** the breath of life for all that lives. They are the ebb and flow of the two aspects of nature's forces moving within **"The Continuousness",** of primordial substance.
"The Connectedness", facilitates the carrying of, the aspects of energy throughout the cosmic ocean, The two aspects of energy are the swells and

troughs of waves, rising and falling in a vast ocean. The aspects of energy form the "fields of force", that hold the chemical elements together, or form three dimensional resonant standing waves, the electron is an example of this.
The wave fields are of "stresses" due to motion, and "strains", held in displaced substance. **There is no "action at a distance",** in this theory. **All substance is infused with energy**, kinetic energy, and potential energy. These aspects of energy drive "random motion", and also the directedness of "on-going motion, striving to reach a goal.

I mention the following for clarity of meaning, not to stop you from using your favorite terminology, In this book the ancient term **"Soul",** is the third basic component of the cosmic ocean, it is used in meta-physical writings, and by myth makers.
The term, **"Consciousness",** is understood by psychologists, medical doctors, and the scientific community in general when speaking of the activities of the mind. **Consciousness** is also the transporting of the aspects of force within the substance of the entire cosmic ocean.

The Soul
carrier of the life force

The Soul, The Soul, the eternal Soul. Ancient Egyptians at the time of **the pharaoh Unas,** in the 5th dynasty of the old kingdom about 2350 B.C. believed the soul of a person is immortal. **"The Soul", "The Ba",** is that which makes a person live. **The Soul** survives the death of the body. The

Egyptian hieroglyph for the soul is a bird with a human head. The "personality". **"The Ka"** is the behavior of a person, the Egyptian hieroglyph for the "Ka" is a stork with a human head. The Ka is not immortal.
Hindus believe the soul is immortal, they term it **"The Atman" The Self**, and describe it as **"The True Self".** In this book **The Soul** is nature's forces, The Soul is the messenger of the Will of the god's, and men.

In this book, The Soul, The Consciousness, and The Connectedness, are the same, They are the forces of life, energy's message, transported within the continuum of substance, transported by "The Connectedness", of substance. **Aristotle** (384-322 B.C. Greek philosopher), saw **The Soul** as being of three aspects or types, naming them, the **Divine Soul, Human Soul, and Animal Soul**. psychologists saw three aspects to the mind, **The Subconscious, Waking Conscious, and The Id**. Scientists see three aspect to the flow of the forces of nature (displacement of substance, velocity of substance, and acceleration of substance). The forces of **Action,** and the forces of **Reaction.** Each strive for a dynamic balance, using a, **Cycle of Change**.

Some believe a person has two aspects to the soul, they see a higher and a lower aspect.

> **Higher Soul**, a soul of potentialities, reason, ideals, and memory. These are also qualities of the higher mind.

Lower Soul, a soul of actualities, on-going constant needs, and desires. These are also qualities of the lower mind.

Now this is all well and good, some of these terms and ideas were formulated fifteen hundred years ago, others are somewhat more recent. The persons that wrote them down did well considering the information they had at the time, but today those ideas can be refined a bit. In the theory in this book, the activity of the soul, the carrier of the power, the forces of life, may be that of a person or the entire universe. **The Soul's ability to carry the forces of nature is due to the continuousness of primordial substance.** The Soul transports the aspects energy and as a result also their messages. The Soul carries the power that animates the body and the mind. The Soul is the carrier of thought and the transporter "The Life Force", "The Breath of Life". in this book, "The Child of Duality", "The Consciousness", and "The Connectedness", are the same as the Soul.

In **Fig. 6** below, in the squares of the top horizontal row are shown things with **Potential Aspects**, In the bottom horizontal row are shown things with **Actualized Aspects**. Both these rows contain the three basic components of the cosmic ocean.

The primary gender of the aspects in the top row is feminine, this top row has the squares with primordial substance, potential Energy and Counterforce. In the bottom horizontal row of **Fig. 6**, are shown the complementary opposites of the things in the top row. The primary gender of things in the bottom row are masculine and include, Time, Kinetic Energy, and actualized on-going Forces of nature. Energy is continually on-going in it's animation of substance. Substance is the mother of all things, Energy (the non-substance) is the father of all things. .Efforts to bring change using kinetic energy, are met with a reaction from potential energy, to hold onto conditions already present.

The secondary aspect, (the Shadow aspect), of things is of the opposite gender to that of their primary aspect.

Notes

	P rimo rdial Substance	The "Link", Between Substance and Energy.	The Non-Substance Energy
This Row the Diagram is for that with Potentiality and is of feminine gender and reactive aspect.	Space it's primary gender or aspect is feminine. The mother of all things.	Counter-Force it's primary gender is feminine. a reaction force.	Potential Energy it's primary gender or aspect is feminine. Energy held as strain forces in substance.
Tish Row of the Diagram is of either gender or aspect. it is a "Striving", for "Dynamice Equality" "Dynamic Balance". Depending on Conditions.	Chance - Randomness regulated by a C ysle of meetings and misses.	The inevitable progression of a cycle. The Cycle of Change.	A sharing of Existence. Alternation f the aspects or genders of energy. Choice - Selectiveness
This Row of the Diagram is for that which is "Actualized", an on-going Actuality.	Time, it's primargy gender or aspect is masculine. Time lets things "happen". Father Time.	On-going velocity of substance. This "Effort" force to bring change has as it's primargy gender or aspect "maleness". The stress forces that that motion can exert on substance.	Kinetic Energy, the Energy possesed by Moving substance. it's primary gender or aspect is masculine. The animator, father of all things.

FIG._6_ Genders of the aspects of the basic components,
Placed in a Matrix Diagram

The Planes or Worlds of Existence are not shown in the above matrix diagram Fig. 6 , to make the diagram more easily understood.
If the "planes", or "worlds", had been shown, they would be:

Atma the three squares of the top "Potentiality", row,

Kama the three squares of the bottom "Actuality", row,

Buddhi the three squares of the vertical energy column, and

Sthula, the three squares of the vertical Substance column.

The world of Mind is a rather distributed affair, distributed over the whole of substance. Substance acts like an orb spiders web sensing all change in the substance, Energy is infused in substance and animates the mind. **Atma** takes care of memory, understanding, and reason functions, **Buddhi** takes care of logic functions, selections and choices. **Kama** takes care of sensory functions, and **Sthula** takes care of "fields", and "forms" of things.
In the vertical central column between substance and non-substance (energy), is that which lets substance and energy communicate are the **forces of nature**. They let the two extremes, substance and energy interact. In this central column you will find the forces of change and the counter forces reacting to change.

Note: in the theory in this book, both **"Space"**, and **"Time", are "Substance"**, and so are in the vertical substance column of the matrix diagram.

I have been wittering-on, explaining that is, about the basic component's of the cosmic ocean, long enough. Let us now look into the "means", of regulation, the "means", of controlling reality.

Part Two

The Heart of Reality

The Regulation of change in the Cosmic Ocean

Man has pondered for eons what regulates and controls events, and people's lives, pondered what controls destiny. The destiny of things or persons begins from simple causes, and then all to soon can get complicated.

An example illustrating that life is not always as simple as some would like it to be, is the speech Dr. Martin Luther King Jr. gave a on August 28, 1963, before a crowd of thousands in Washington D. C. he had a three page prepared, typewritten script, after the speech a last minute security volunteer, 26 year old George Raveling, asked Dr. King Jr. if he could have the paper copy of the speech he was still holding, and Dr. King handed it to him., It is thanks to George Raveling it still exists today and it's words can be re-read. The thing is, no where in it does it say "I have a dream", also the speech was untitled. What it did have was an asterisk mark to indicate Dr. King could add lib at that point, to extend the speech if he wished. Martin Luther King Junior's ad lib began with, **"I have a dream"**, and the adlib lifted the speech out of the ordinary, and into the realm of the extraordinary. George Raveling almost did not go to hear the speech on that day, Fate guided events.

As in the above example of "Chance", randomness, and coincidences, unexpectedly alter our lives and our world. On the cruise ship "Oceanus", off the coast of South Africa, two passenger entertainers, Moss Hills and his wife Tracy, were entertaining some passengers, when Moss sensed something seemed amiss with ship. The ship had begun rolling an extraordinary amount he believed and no one had explained to anyone why, or what to do. Moss went down into the bowels of the ship to see for himself what was happening down there, it was flooding with sea water, is what it was doing, so Moss went up to the ship's pilot house to tell the captain, but there was no one in the pilot house. Moss then radioed for emergency help, it was fortunate he knew how. His call reached a South African navel station, and he explained what was happening, they asked Moss "What rank are you?", and he replied "Guitarist". I thought that was a wonderful response. The naval people then wanted to talk to a ship officer, and Moss replied "There are none in the pilot house, it is empty." he probably added "Hurry I do not think we will stay afloat long". Rescue helicopters came rather quickly, the six hundred passengers, and the Hills, were saved. The ship "Oceanus", sank almost immediately after the rescue. It was chance that Moss Hills had acted as he did, and that the six hundred passengers had booked on that same voyage with the only person that had radioed for help. You can actually see Moss Hills, pictures of the rescue, and find out about the empty pilot house, simply view the "Nova" TV

program titled "Events leading up to Cruise Disasters", these were actual events, not fiction.
Chance intervenes in one's life perhaps more often than thought, sometimes with greater effect than at others. To celebrate my graduation from high school (12th grade), my parents, brothers and I went to a well known restaurant in town. As was the owner's habit he would stroll around and chat with some of his customers and ask "How are things going?", and so forth, on finding out it was my graduation and as I was of military conscription age, a memory from his own youth may have been triggered in his mind, and he began to relate an experience he had while in the German army. One day on the battle field he came upon a helpless wounded German officer lying on the ground, he helped the fellow stand, then carried him to safety and medical attention. He thought no more of the matter, the months went by, a day came when he was told peace has been declared, "you can go home". So he began to walk the eighty miles or more home, on the way he came across another German army unit that had not heard peace had been declared, they grabbed hold of him and in spite of his claim that the war was over, he was taken before a table set up in a grassy field and charged with cowardice and desertion. Both justice, and injustice can be swift in a war zone, if found guilty he could be shot by firing squad. Then one of the officers setting at that "field court marshal", table, looked him over, said "I know this man he is no coward, he saved my life on the battlefield at risk to his own life". It was the officer he had carried off the battlefield, what are the odds they should ever meet again? The verdict was not guilty,

and he was freed to go home, eventually he immigrated to America, and became the owner of his own restaurant, and there I was with him hearing his tale in person, all quite by chance.

Hindu thought on controlling change

Hindu's have personified the regulating aspects of the lives of all living things, as being accomplished by a trinity of gods, the trimurti,, their names are:

Brahma, the god that brings order to the universe, the creator of "things born". (**Choice**, the selection of what is to be done.

Vishnu, the sustainer god. The god that keeps the flow of change going, **The Cycle of Change**, as regulator.

Shiva, the god of dissolution. The destroyer that makes way for the new. **Chance**, the regulator of the random time, and place, an event occurs.
Who can say what place, century, year, hour or instant, a certain event will occur, a god of randomness could, chance itself.

The Hindu term "Buddhi", is understood to mean two things, one distinct, the other more murky. The distinct meaning of Buddhi is, the animation of substance by energy, and that is the view taken in this book, because in the theory in this book all substance is animated by energy, mobilized by energy, chaos is driven by energy. It is necessary to employ energy to move substance in the control of change as in an electronic computer or the human brain.

A second more murky meaning of Buddhi, is the regulation, or control, of energy. This second meaning I have treated separately as there are the three "means", or methods used to regulate a single outcome. Arriving at this single outcome is the task of what I have termed "The Heart of Reality". The Heart of Reality, regulates the "flow", of all change in the cosmic ocean, it does so using the three simple "means", In the western world the "means", have been personified as "The Three Fates".

Fate

Fate is an impersonal force that predetermines all events, it is the cause of destiny. The spoken decree of the gods, the will of the gods.

Some would say, the cause of destiny is **Need** (necessity), others say it is **Luck** (chance), or still other say it is the selections we make, as the **Choices** made, that alter the coarse of one' life. In a way this last cause of destiny, might well be the power of **Love** (a power that has altered many lives). The ancient Greek philosophers had their own ideas about "Fate".

The Fates, Hold the Threads of Destiny,
or put more picturesquely,
The Fates weave the cloth of one's life.

In myth, the personified **"means",** of regulating the coarse of events, or the behavior, of the three components of the cosmic ocean, has been called **"The Fates"**.

Other names for personification of the regulators of destiny, are the three **"Norns"**, to still other persons Fate is **"The Will", of Goddesses and Gods**.

Let us first look into the word Fate. Examples of the word "Fate", in different languages are, **wurd** (low German), **urdr** (old Norse), and **wyrd** (Anglo Saxon). In Shakespeare's play "Macbeth", there are three wyrd sisters. **Sisters of Fate,** sisters of destiny. "Double, double, toil and trouble; fire burn and cauldron bubble." (Act 4, lines 10 and 11). Probably, to be politically and religiously safe, in that day and age, when the play was written, these sisters of fate, were termed "witches" to suppress an older belief, that the sisters used magic and spells to get their results.

The three sisters of Fate were personified **as "The Moirai".** Some described them as old wise women, and from the time of **Hesoid** (800 B.C.-Greek poet) onward, were in myth termed, **"The daughter's of the Night"**.

The Fates are seen as, the personified **"Means",** or method that regulates a person's life, and that of all that lives.

Lachesis - **Chance**, Luck, The meeting of the random uncertainties of existence. Gamblers use probability to predict results.

Probability, is the introduction of "odds" of an occurrence in the field of pure chance, like the odds of rolling a certain number in a dice game, such as can result in a game of "Craps".

Atropose - **Certainty,** Inescapable fate, against which there was no appeal.. For a person the most absolute certainty is their birth, and death. Here "The Inevitable", is a cycle of existence between a thing's beginning and it's end, a life-span.

Clotho - **Choice**. The power to choose one's destiny, by making choices. Being one's own weaver of the cloth of life.

In Norse and Teutonic Myth they are **Norns**. The dispensers of destiny.
Astrologers predicted a person's destiny, their fate, by the position of stars and planets at the time of the person's birth, to which **Shakespeare** in his play "Julius Caesar" wrote "The fault dear Brutus, is not in our stars, but in ourselves. " Act 1, scene 2, lines 139 to 141.

The Three Norns, Mistresses of Fate (Destiny), Female Spirits. seen as **Experience**, past, experience present, and future experience.

Urd (old Norse), Wyrd (Anglo Saxon), the wisest of the Norns.
Here, **"Past Experience"**, What once was, Change that once had been actualized.
Verdandi (old Norse), Here seen as: **"Present Experience"**,

What is, experiences of the "Now", can be said to be, "true", things, fact. Perceptions of experience are sometime partial fact only, sometimes imagined fact.

Skuld (old Norse), Here seen as: **"Future Experience"**. Experience that may come to be. Potential change.

Alternate spellings of the three Norns: **Urth**, **Verthandi,** and **Skuld.**

The matrix diagram of Fig._7_ shows the placement of the fates as they relate to one another and other aspects of the diagram. These Determiners of the Course of events, as **"The Moirai".** occupy **the middle square of each vertical column** The terms that interact in a vertical column are in the squares above and below the middle squares. The middle squares are the result of the interaction.

You will find on looking at the matrix diagram Fig. 7 , that **Chance** is the regulator of **Substance** and **Choice** is the regulator of **Energy**, **"Chance",** is the coming about of Unplanned encounters, things or people come together by accident and affect one another. Randomness "happens", Unplanned encounters "happen",

"Choice", **occurs in two general ways, a)** - Complementary opposite features encourage a joining together (unification) to "Make a whole", Complementary opposite gender persons can be attracted to one another and become as one. Oppositely charged particles are attracted to one another and may become as one. Things can be joined by reason, as when two things are recognized as forming the halves of a complementary opposite pair.

b) - Non complementary opposite features discourage attraction, the difference separates them, "discriminates", them, as if a wedge were driven between them, repels them apart. Logic is a mechanism to help arrive at a decision or select a way to proceed, a way to choose a "yes", or a "no", "to start", or "not to start", "Like", poles of magnets or "Like", charged particles "choose" by repulsion, by pushing one another apart.

A Cycle of Change, regulates the forces of nature. This cycle has three characteristics: 1) Displacement of substance, 2) Velocity of displacement, and 3) Acceleration, a change in the velocity of displacement. All things undergo this cycle of change, it is a continual communication between substance and energy.

Placement of The Three Regulators, the Fates, and the Norns, in a matrix diagram.

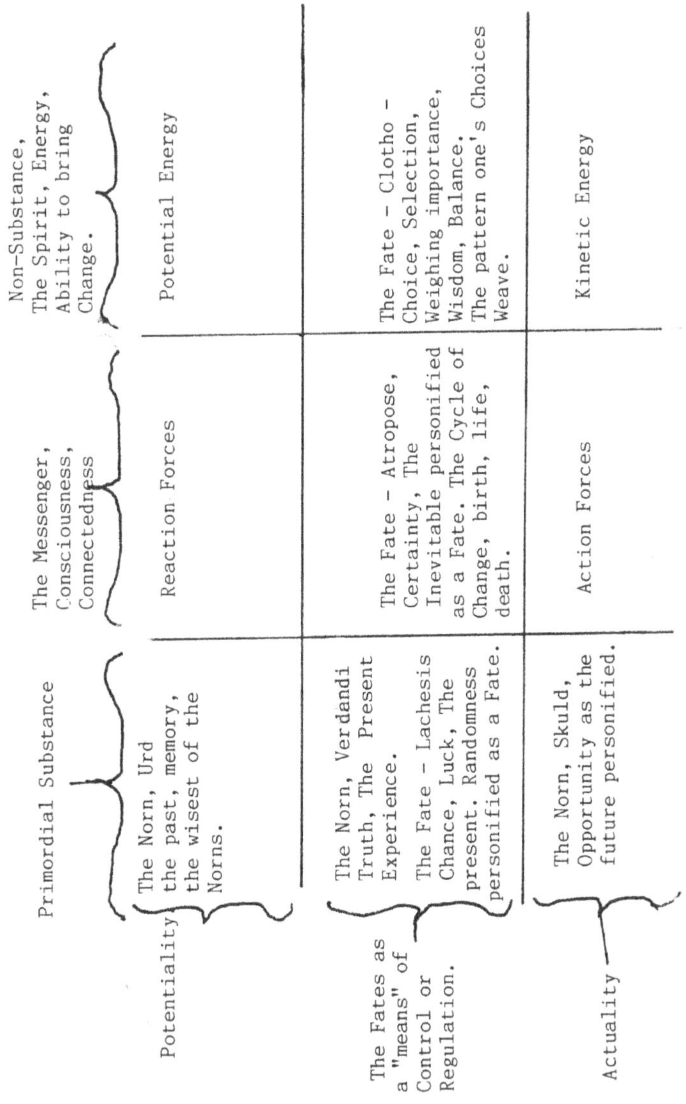

FIG._7_ The Fates and Norns, Mistresses of Destiny
The Regulation of the two Aspects of each Basic Component of the cosmic ocean, and the Goal of that Regulation,

To make clear what is meant,
a brief review is now presented:

The three basic components of the cosmic ocean, are Substance, Non-Substance (energy), and that which connects them, the oneness of substance, and "Force". The two complementary opposite aspects of each of these basic components are as follows. The aspects of substance are **"Space"**, and **"Time"**. The aspects of The non-substance (energy), are, **Potential Energy**, energy held in substance and **Kinetic Energy**. the energy possessed by moving substance The aspects of force are, the actualized effort force to bring change, and the potential counter-force to hold things as they are.
The Fates are the personified form of the three "means", used to bring about regulation. The Non-Personified names of these regulating "means", are:

Chance - regulates the behavior due to the aspects of Substance.
Choice (also termed Selection) - regulates behavior due to the aspects of non-substance (energy).
and
The Cycle of Change, - this regulates the ebb and flow of nature's forces, regulates the action and reaction to change, it's "Forces", and "Counter-Forces". This Cycle regulates the characteristics of the transfer of power between substance and non-substance, This **Cycle of Change**, has three characteristics, they are **1)** The **Displacement** of substance,

2) The **Velocity** of substance, and

3) The **Acceleration** of substance (a change in the velocity rate).

Does life have a purpose, a goal?

The world at times seems rather chaotic, and the actions of others are not always predictable. A noted scholar of mythology once commented that, life was a bunch of protoplasm with **an urge to reproduce and continue in being**, and added to this "and have a bit of fun".

Men and women have many different purposes as they live. If I interpret correctly the sentiment of French author Antoine de Saint-Exupery (1900-1944) expressed in his book "The Wisdom of the Sands", (French title "Citadelle", that men and women **need a goal, a purpose, to strive for as they live,** a purpose, that is always just out of reach, it is like hearing a siren call from an island of dreams, a reason to go on striving to reach the goal, Examples of this are: (for a man) **longing to be with a woman when being away from her.** Or a pearl diver, diving to the ocean's bottom to find a valuable pearl, even though it is dangerous, or a sculptor, or painter striving to create, a beautiful work of art, a poet striving to create a poem that sings to it's readers, a carpenter striving to turn a pile of lumber into a ship capable of traveling the seas, or a needed shelter. Once the desired woman or object is possessed the challenge is gone, the longing for it is gone, the drive to achieve it, is gone.

I believe what author Antoine de Saint Exupery was getting at is **the striving has more value to the doer, than possession of the object, the longing or striving gave a meaning to their life**. Once the goal is accomplished a new goal will be striven for. Others have expressed a related sentiment as

"Wanting is not the same as having. His other comment was a man needs to earn the love of a woman, simply paying a fee for permission to use a person, is not "earning". Let us now see that, "striving", for something, is only a part of regulation, it's other part is achieving the goal, achieving satisfaction. Each affects the other.

The Goal of Regulation

The goal of, "The Heart of Reality"

It took me time, months actually to realize the three regulators had to have a goal a purpose, and that the "means", of each was different, they had a common goal, a common purpose to carry-out. Laugh if you will, but to me at first glance it seemed the three "means", controlled such different things, as "Behavior", of substance, "Changing States", of energy. The inevitable Cycle of Change, seemed to be regulate it's own three characteristics, (displacement of substance, velocity of substance, and change to velocity of substance). So did all three have a common goal? Then it dawned on me, **the goal, the task, of each "means"** of regulation that was **continually to striven for was "Dynamic Equality",** to strive for a shifting changing goal, a dynamic goal, a "dynamic", equality for each extreme aspect of it's duality. Each of the three basic components in the cosmic ocean, has two extreme aspects to it's nature, each is a duality.
For example there is a Dynamic Equality striven for, between Space and Time, in the substance category. A Dynamic Equality striven for, between potential "held", energy in substance, and kinetic energy,

energy of "moving substance", in the energy category, and finally a Dynamic Equality striven for between the Forces of nature, it's duality is the forces for change, and the counter forces potentially keeping things as they are.

Force, lets substance and non substance communicate.

The three determiners of destiny, that Strive for Dynamic Equality

Chance, In the realm of Substance, restless energized Substance, brings about random encounters and collisions of substance with substance.

Choice, operates in the realm of Energy, Selection leads to realizing a pair of complementary opposites are halves of a whole. Logic can give direction to change.

Cycle of Change, the continual flow (interaction) of force and counterforce, striving for a measure of "Dynamic Equality", between extremes.

These "means", to "Dynamic Equality", "Dynamic Balance", continually strive to level the playing field, they are also valid in various fields of science, such as, physics, where the striving is between the states of chaos and order in the cosmic ocean. In mechanics where the striving is between the force and counterforce, and, In biology where the striving is to achieve "Dynamic Balance", between species, so each species may survive.

In the efforts to achieve the ever moving goal of "Dynamic Balance", the cycle of change is repeated, and repeated and repeated.

It is important to understand **"Static Equality" is not** what is to be striven for, the reason being energy is bonded to substance, and all substance is jiggling, and in motion, due to energy animating it, the whole system is in a state of continual "Change", a state of "Flow", a state of "Flux", "never still",
The Energy's aspects keep adjusting, bringing "more", change here, and less change there, as needed. It is not a question of trying to stop change, energy is eternal and is always present it is a matter of which aspect predominates, potential energy stabilizes and holds with strain forces in substance, kinetic energy is possessed, by substance in motion, animating it.
"Dynamic Equality", is what is striven for as a purpose, a goal, it is a moving goal that will never be reached, it is always ahead beckoning, like the desires of an individual. The system continually swings in a cycle from one extreme to the other, destined to be eternally seeking that which it has not yet achieved. There will always be, space and time, there will always be, the potential and the actual, always be force and counter force, it is just a matter of which dominates in a situation. The pendulum of "Dynamic Equality", swings eternally from one to the other, never still. The pendulum's swinging is a seeking, a striving for "Dynamic Balance", it is the beating of the human heart, the pulse of nature, the cycling of "The Heart of Reality".

Descriptions of the Dynamic Equalizers"

1) Chance, is Luck, **is Randomness, is Freedom**. Chance is what choice is not, in that they are complementary opposites.

It has also been said that Chance is, "being at the right place at the right time", or the "We met quite accidentally", type occurrence. Random motion, random collisions, Lady luck, **"Chance", is the unpredictable companion of all men, and all things.** Chance is the regulator of substance. Chance can be described as **"The Masses Yearning to Breath Free",** whether atoms, or people.

An example of Chance in operation, is in a game of dice, the numbers facing upward after a roll of the dice are the product of Chance in an honest game. Chance, does not deny that the Choices a person makes also affect one's life, one's destiny, **it however claims**, that chance, due to random events occurring at the same time, such as random coincidence, sickness, or accident, strongly influence the events in one's life.

Example: If a person chanced to be at a robbery scene, and were held hostage or wounded, it may influence their future.

2) Choice, is **Selection, is Limitation**. Choice is what chance is not, Chance and Choice, are complementary opposites. Choice is a directed purpose, a selected direction for effort to proceed in. Selection can match together complementary opposites, or it can operate like the right key to unlock a barrier and let a process proceed a step further, repelling some tendencies and attracting others together. Bringing about a tendency **"to be", whole,** or **"to be", individual as** one's self.

3) The Cycle, is **the Inevitable path of Change**. A fateful process like, the cycle of life running it's natural coarse, with but one beginning, and one inevitable end. A person is born, grows, ages, declines and falls into dissolution, no longer able to act as a whole, and fades back into the cosmic ocean background, from whence all originated. The bob of a pendulum swings back and forth in wave motion, continually accelerating and decelerating during it's wave cycle. **A Cycle, with three characteristics**, as do all cycles, the characteristics are **Displacement of Substance, Acceleration of substance, and Velocity of substance**. **Please note: "acceleration",** means an increase in velocity, but can also when present, refer to a decrease of velocity (commonly referred to as a **"deceleration"**). **The Cycle,** is a striving for dynamic balance. in the world of science a cycle has a positive half and a negative half, an alternation of increasing and decreasing displacement. In the instance of **a torsion wave**, it's displacement's twist is first in one direction (clockwise), then in the in reverse direction (counterclockwise). Each half cycle is of equal duration.

	Primordial Substance	The Forces of Change. Effort.	Non-Substance Energy Animation Ability
Having the Potential To Change	That which can have "form" Space	Acceleration of Displacement Reaction Forces	"To Be" Complete Attraction Selection of Complementary Pairs
The "Means" of Regulation	Chance, Luck Freedom of Substance Substance Behavior	Equality The Inevitable Fate, The Cycle of Change.	"To Share" Dominence. Alternation Choice, Limitation
Actualized on-going Change	Time Sequence of existence. Duration	Velocity of Cisplacement of Substance on-going Action Forces	"To Be" Individual Repulstion Implementing Direction Change is taking.

FIG. 8 The three regulators of the Coarse of Events

They are the "means" of directing the path, or direction, of change.

To review the "Inevitable" cycle of fate's three characteristics.

Displacement of Substance - displacement is the moving of substance from a reference position, to a different position. Displacement may be a longitudinal motion, or a twist, a partial rotation, a torsion motion.

Velocity - Velocity of substance displacement. Velocity is an on-going actuality, not a potential thing.

Acceleration - a change in velocity of substance, a change in the rate of displacement of substance, this is a reaction to change, and therefore a potentiality, not an actuality, until it happens.

The cycle of change varies the amplitudes of three characteristics, during a cycle of change., see graph Fig._21_ page 138, for comparison of their rising and falling. In terms of human experience, the three characteristics of the cycle are:

Need - is ongoing force, a force demanding change. need is an on-going actuality, on-going velocity.

Satisfaction - is the minimizing of need. Satisfaction, is a potentiality, it may or may not occur. Satisfaction, is a change in the rate a thing is occurring at. and

Emotion - is like a displacement of the substance of thought. Emotion is a striving for a state of resolution, between, the forces and counter-forces of thought, a striving of "need", to achieve "satisfaction", of that need. This is a cycle of "striving", for dynamic equality.

Just speculating now, If we picture Will and Wisdom as the two extreme aspects of energy, with **"Will"**, as the "potential", to choose, **"Action"**, as the "actualized", choice, Then finally picture a sharing process as an alternation, a striving between "holding on to", what one has, and a "loosing", of what one can loose, for the better. The alternation becomes one of repeated comparison, and a choosing, a choosing of which aspect of power shall predominate, and which shall simultaneously become subservient, this process, has been is termed **"Wisdom"**. With these three, we have the metaphysical, **Will, Wisdom, Action** of metaphysics. It is then understandable why the ancient thinkers, mixed regulation of energy in with energy as animation. The terms they gave this mixture, **"The Logos"**, and **the "plane of "Buddhi"**. is represented in the vertical column of "energy", in the matrix diagram (Fig.__8__).

Will - is the "held", power of what once was or can be. A thing, condition, or a hoped for vision of an existence as one would like to be, a possible goal, a potentiality, it is **(Dynamis)** , a held state, **Will**, has the characteristic of, the Guna Tamas.

Wisdom - "The way". The way of sharing simultaneous existence, by alternating as thought necessary, which of two complementary opposite aspects of energy **becomes** predominant as simultaneously it's opposite aspect becomes subservient. **Wisdom,** has the characteristic of harmony, the Guna, Sattva.

Action - A decision to actualize a choice, to give change a direction. Also action is a separating, a discrimination, or a judgment, an actuality, **(Energeia)**. **Action** has the characteristic of, the Guna Rajas.

Has man put the three regulators of the cosmic ocean to practical use?

Answer: Yes, From **Chance** we get gambling dens, and the study of "Probability". Also insurance company "risk", classifications.

From **Choice** we get sports betting, where the betters "choose", the potential winner, and place their bet accordingly, also we get the science of logic. Electronic computer calculations are based on logic, it uses a series of on-off switches termed "logic gates", to route a signal much like a person saying "If this happens, I will do that, and if the result is "such and such", I will do this."

From **the Cycles** of nature, we get a measure of certainty, and repeatability, in our lives, and the study of dynamics, the rise and fall of the three characteristics of a cycle of change, (velocity, acceleration, and displacement). Examples of repeatable events, are the seasons, a woman's menstruation cycle, referred to as "the curse", by some, and the swing of a pendulum. The cycles of nature awaken a sense of wonder and mystery in those that experience them. The buzz of a bee, the chirp of a cricket, the smell of a rose, the rolling sea, a mirage over a heated road, electromagnetic waves, and the cycling of beating wings. An example of this fascinating phenomena, that can not only be seen but heard as well follows.

It was on a warm summer eve, that I stood watching one of nature's marvels in action, a Hummingbird Moth (Hyles lineate) darting over a bed of red petunias, selecting a flower, and then hovering in mid air above it, to sip it's nectar, what a privilege it was to be there with the creature. A nearby park had two, three foot wide by fifty foot long, petunia flower beds, where if one were fortunate, just before dark, three or four of these little fellows would be searching for nectar, On observing them from two or three feet, hovering in mid air their wing beats could be heard as a just audible low Woooooooo, at 85 beats per second. (A true humming bird's wings beat faster at about 200 wing beats per second in normal flight or hovering). There is more than one species of the insect popularly known as the Hummingbird Moth, examples are: **The Sphinx Moth** (Celerio lineate), **The white lined Sphinx Moth** (Hyles lineate), **The pink-spotted Hawk Moth** (Agrius cingulata) and others.

Potentiality (Dynamis), **and Actuality** (Energeia),

Dynamis can be understood as "held things", and Energeia as "the on-going velocity of things". Between them is their, their child, a repeating **Cycle of Change, a cycle of striving** to achieve "Dynamic Equality", striving for Dynamic Balance, balance over time.

Two pairs not listed are Real and Unreal, and Existence and Non-Existence. These were left off the list purposely because in the theory in this book, **all that exits is real. Unreality is non-existence.**
Actuality, is on-going efforts to change that which exists. **Potentiality**, is that which exists, is possible or imagined.
Mathematicians were wise enough for their purposes to have "real", and "imaginary", numbers", at least they did not term any numbers "unreal".

Some Chinese mythologists at the time of the Chinese emperors, saw all things as "active" (male) or "passive" (female). Psychologists of today believe a person's mental picture of their "Self" has a duality to it. **"The Self" a person believes they are**, and a complementary opposite self, a "Twin Self" of opposite sexuality, a "Shadow Self". This Shadow Self is **"The Self", a person believes they are not**. Folk tales of Ancient Egyptians told of a person's twin self, a self of gender opposite to the gender one is physically born with. This twin self as the tales goes, exists elsewhere, in another land. **"The Mind", is another land**, a land in which each person can roam and meet their "twin Self", and more, even "be" their twin Self.

Some myths give the powers of nature a human or animal form, they Personify them, and term them goddesses, and gods. Examples are the Teutonic god of light **Heimdall**, and the Scandinavian god of light **Balder,** the Teutonic god of thunder **Donar**, the Scandinavian god of thunder **Thor,** also **Woden,** or **Odin,** principle god of Teutonic peoples, the mythical Woden, master of "fury" had the magical power to incite panic in the enemy. He was the leader of a band of phantom warriors, ghost warriors who on stormy nights could be heard galloping through the clouds in the darkened sky.

Days of the week have been given the names of gods, and heavenly bodies.

Sunday (Sun day)
Monday (Moon day)
Tuesday Tiwes day after the Germanic god Tiw or Tyr. (Mars)
Wednesday, Woden's day, after the god of Teutonic peoples, Woden, (Mercury)
Thursday, Thor's day after the god Thor, the Germanic Donar. (Jupiter)

Friday after the Norse name Frigg, old German Frija, or Frya. (Venus)
Saturday, Saturn's day, after the Roman god, Saturn. (Saturn)

The Five ancient "Elements"

(listed from lightest to heaviest), here they are compared to Metaphysical levels of being,

1) - (Atma) **Aether** At one time the ancients thought Aether was a substance through which the planets moved without or very little resistance to their motion.
2) - (Buddhi) **Fire**, The world of Energy, animator of Substance.
3) - (Manas) **Air**, The world of thought. In one way, The Mind, Manas, is the meeting place of the human and divine. The meeting place of Potentiality and Actuality. To some the Earthly meeting place of Heaven and Hell.
4) - (Kama) **Water**, The Swamp of need and desire, a quagmire one is continually involved with. Kama is, on-going flow of efforts to bring change. Velocity.
5) - (Sthula) **Earth**, Experience of the moment, fact. The behavior of substance acted upon by energy.
6) - (Paran-Atma) There is no ancient element for this plane. Numbered here as zero. Paran-Atma is all planes that may occur or have occurred. Paran-atma is eternal existence, formed or unformed, with or

without behavior. Paran-atma is Brahman, and is, beyond words.

The Supports of the Cosmic Ocean

Long ago it was believed, the cosmic ocean had four supports, as if the sky were a roof, held up by four posts.
Aristotle wrote of the four causes of things. (Efficient cause, Material cause, Formal cause, and Final cause.
The Aryan Hindus spoke of the "Worlds" or "Planes of existence". There are five or six "Worlds" (depending on which a person cares to include in the count). I have included four of these as part of the basis of the matrix diagram arrangement, these four are:

a) - **Atma,** in Atma, is the Potential for Change.
b) - **Kama,** in Kama is Actualized on-going Change.
c) - **Buddh**i, in Buddhi is Power, Energy, the animator of substance, and it's regulation.
d) - **Sthula**, in Sthula is Primordial substance, and it's behavior.

The Biblical Genesis story speaks of man being made in **the "Image",** and **the "Likeness"** of God.
The **Image** part, is the forms, the shapes, primordial substance takes on under the influence of energy. The "Image", is the body of reality. We primordial substance to the clay a sculptor uses to mold into forms.
The "**Likeness",** part, is the power, the **ability to bring change.,** the ability to animate substance, to as it were bring the clay to life.

Dynamically Unbalanced Conditions keep the cosmic ocean in a continual state of change

Unbalanced conditions are **the engine of change**. As long as a duality exists, a state of imbalance exists. Each extreme of the duality longs to merge with it's partner, yet it also longs to be an individual free to live it's own life. Obviously each cannot do both "full time", they must share existence, by alternating which is master at any one time. There is a continual striving for "Dynamic Equality", "Dynamic Balance", between the two extremes.

The existence of "Duality", maintains the imbalanced condition. Pairs of complementary opposites, have a natural tension between them, or else they would be attracted to one another and merge, unifying back into the unity from which they originated, returning back into their common source, a cosmic ocean without dualities.

A few pairs of dualities follow, others can be found on pages 22-23.

Actuality and Potentiality
Substance and non-substance (energy)
Self, and Shadow Self
Self and Other (not self)
Body and Mind
Cosmos (order) **and Chaos** (disorder)

To exist as dualities, and avoid joining together into a "oneness", the two aspects of duality each alternate between a maximum amplitude and a minimum amplitude, first one becomes "dominant", as the other is becomes "submissive", then they alternate the roles and the submissive one becomes dominant and the dominant one becomes submissive, This is done to achieve a "Dynamic Balance", so both aspects of duality can co-exist without uniting, yet achieve a mutually equivalent existence.

Unbalance fuels this engine of change, the engine operates continuously, "striving", for "Dynamic Balance", at any instant, it's goal is just beyond reach, the engine is not discouraged by this, and in the next instant, or next cycle, **maintains it's quest for, that measure of "Dynamic Balance".**

"The Connectedness", "The Consciousness".

The Continuum of substance, Carrier of the forces of nature.,

Nature's forces give form and character to primordial substance, and they are "the breath of life", within substance. but it is the oneness of primordial substance, it's **"connectedness"**, that lets the force travel within and affect all the worlds. Conduction of force gives substance an awareness of conditions within substance, this keeps the whole "in touch", with it's parts.

Note: the term, **"Consciousness",** is favored by psychologists, medical doctors, and the scientific community when speaking of the mind's awareness of the conditions of existence.

The term **"Continuum"**, is favored by scientists and science fiction writers, when speaking of, **the uninterrupted extent of a thing**, it's oneness without voids or breaks. The body of the cosmic ocean is a "continuum", facilitating the conduction of change. The aspects of power move about everywhere in the space continuum's uninterrupted extent.

The effect of force on substance results in **Fields of Force,** these are the strong arms of change. They hold substance in resonant standing waves such as those of an electron, and a unique group of torsion wave fields form each chemical element.

On looking over Figure__9__ you will see the forces are in pairs of complementary opposites, what one is the other is not. The top square has the **"strain forces"**, these are **"reaction forces"**, these are due to deformation or displacement of substance. These are the counter forces any effort to bring change encounters.

The bottom square has forces that, are on-going directed efforts to bring change, these are **"stress forces", kinetic forces,** forces directed to drive substance into deformation or displacement.

The center square of Fig. 9, represents where the forces are resolved. This is where "Push", the bringer of change meets "Shove". Shove is counterforce holding the shape of forms, like our own forms.

"Consciousness", is an awareness in the mind of conditions it is exposed to. In Fig. #6, or Fig. 9, the center square, represents where force and counterforce are resolved. In the mind the resolution is expressed as emotion.

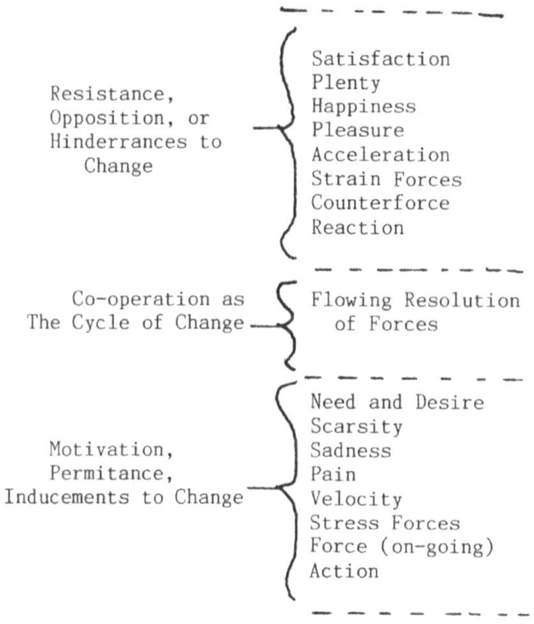

FIG._9_ The Flow of Effort (Force).
"Forces", encourage or discourage deeds.

Since change occurs over a period of time I believe perceiving it as a form of wave motion, is useful, waves facilitate conclusions in fields other than physics, fields such as economics (which has the velocity of money), and economic depression and inflation periods. Psychology has periods of accelerated behavior change, or in the field of thought, logic process and sequence cycles, etc.

The two Aspects of Effort, Force, and Counter-Force.

All change, and all motion of the substance of the Cosmic Ocean is driven by energy, in either of it's aspects potential or kinetic, and posses three characteristics: **displacement** of substance , **velocity** of substance, and **acceleration** of substance**.** These three characteristics are present in not just the motion of substance and materials, but also in the motion of ideas, concepts, availability of employment, changing states of mind, changes in behavior, the changing of the state of the economy, and the transfer of wealth.

An interesting aside is, even money can have "**velocity**", The velocity of money is roughly **the rate** at which a Euro or Dollar note, changes hands in a year, it can also have acceleration of it's "displacement", on moving from person to person as it can change hands quicker sometimes and slower in other instances.

Governments seem to levy a tax on everything they can that the public will not howl to much about (they can always find a reason to levy a tax, once they get a foot in the door, governments are disinclined to take it out and may then expand the tax). Taxing examples are, liquor tax, a sales tax, income tax, property tax, inventory tax, inheritance tax, and a "gift" tax.

In the coarse of a year, one Euro or dollar, may change hands a hundred times, or a thousand times, or more, each time it does so, it is taxed, and can put many times it's face value, into a government's coffers, You may be holding such **a "high velocity",**

Euro or dollar in your hand. How many times in a year, do you think it has changed hands?

Is there a "Still Point", in a Cycle of Change?

Traditionally the still point is defined as an "immoveable spot", or pivot, about which things turn. A pivot such as, the family, the community, even, beliefs held in an individual's mind, are still points about which a person's decisions are made. The center of a sphere is it's "still point", when the sphere is turned.
In the northern hemisphere the month of May is a time of new plant growth after the dead of winter, a May pole is sometimes set-up from which are hung long ribbons, it is a "still point", about which joyous celebrations are held, or children dance merrily in a circle.

The name "Still Point", can be misleading as there are many persons, many families, many communities, many worlds, and many atoms, each is a "still point", about which activity can take place.
Within a "Cycle of Change", such as the life-span of a "born thing", or within one wavelength, of an electromagnetic wave, where is the still point about which activity moves?
I believe the still point of a Cycle of Change, is "The living moment", the moment of the "Now". It is that moment in each of our lives that moves along the band of eternity with our existence, a band comprised of an infinite series of moments of the "Now", each moment a "Now" for an existing

thing. And though the living moment moves through time, it is unmoving to the individual that moves along with it, the individual for which it is the "Now".

The Doggedly Curious, and Science

There are now persons termed "Scientists", who help accumulate knowledge about the natural world, they investigate, classify, and measure factual occurrences. Scientists try to discover and reveal general truths and patterns to change in the natural world. If a person earns their living being a scientist, they are termed a "professional", all other scientists are termed "amateurs". Wonderful are the amateurs for they started the sciences.
Scientists have wondered, speculated, and written "papers", about many things including Heavy water, viscous water, "Poly Water", What is Matter, and "**Is Electromagnetic Radiation, made of Waves or Particles"?**

On investigating the nature of light: There were those who thought light was due to waves, and those who thought Light was due to particles.

Light as waves

Does light consist of waves in a substance, waves of moving energy?
Augustin Jean Fresnel (1788-1827, French physicist) assumed that light was a vibration in an elastic material medium, a medium he termed the luminiferous Aether, the light bearing Aether.**

Further support for a wave theory of light came from amateur physicist **Thomas Young** (1773-1829 English) in 1773, who explained interference of waves of light. **Erasmus Bartholinus** (1625-1698, Danish physicist) discovered polarization and showed light waves must be transverse vibrations**. Then in 1865 **James Clerk Maxwell** (1831-1879, Scottish physicist) showed light to be periodic fluctuations in the strengths of electric and magnetic fields at right angles to one another. In 1887 electromagnetic waves predicted by Maxwell were produced experimentally by
Heinrich Rudolf Hertz (1857-1894, German physicist). **Light is not a substance, light is a process taking place in a substance**. The process can travel through the substance be it "space and time", or for lack of any better substance, scientists of the17th & 18th centuries termed it the Aether. Light and similar radiations can be absorbed, reflected, refracted, polarized, diffracted, even emitted by "Matter". Light is not transmitted instantaneously, it takes time to travel (roughly 186,411 miles per second through the spaces between the stars).

** **Transverse waves** are vibrations that move from side to side, as the wave progresses forward. For comparison **Longitudinal waves**, like sound waves move back and forth as the sound travels forward.

Torsion waves, a third type of waves, are waves of twist, oscillating between clockwise and anti-clockwise rotation, a winding and unwinding activity.

Light as particles

Light may consist of a stream of particles, bits of energy termed "Photons", **Max Karl Ernst Planck**, (1858-1947, German theoretical physicist), suggested that energy is emitted and absorbed in discrete amounts, "packets" of energy he termed **quanta** rather than continuously. Planck assumed that each wavelength had a different magnitude packet. In 1924 **Louis-Victor Pierre Raymond seventh Duke de Broglie** (1892-1987 French physicist) in his PhD thesis, proposed that since light waves could be described as streams of particles, light corpuscles (photons), **perhaps small particles of matter**, such as electrons, protons, or neutrons, **could be described as waves, "Matter Waves".** In 1927 two physicists at Bell Labs in New York U.S.A. Showed that electrons did have a wave nature. Put another way, an electron has a wavelength.

Today "de Broglie's relation" is stated as: a particle of mass **m** and speed **v** will in some experiments display wavelike properties, of wave length equaling Planck's constant divided by momentum, momentum is mass times velocity. A quantum of energy of equals **hv,** and Plank's constant, the constant of proportionality **h,** is **6.547 times 10 to the minus 27th power erg sec.**

Here is the de Broglie relation. **The de Broglie equation.**

Momentum equals Plank's constant divided by wavelength

$$p = \frac{h}{\lambda}$$ lets one determine momentum

$$h = p\lambda$$ lets one determine plank's constant

$$\lambda = \frac{h}{p}$$ lets one determine wavelength

Where **p** represents momentum. **h** represents Plank's constant, and **the Greek letter lambda** represents wavelength. It may be a matter of personal preference and or convenience whether one wishes to see light as waves or as particles, now they can be described mathematically as waves or as particles, and particles can be described mathematically as waves. For a physicist involved in the mathematics of **electron currents** it is convenient to make calculations as a flow of electron particles, whereas a physicist working with energy exchange calculations at atomic and subatomic levels may find it convenient to see electromagnetic waves as **"Wave Packets", Matter Waves"**. Physicists working with light or radio waves prefer working with **"field strengths"** and speak of **"fields of energy"**, electric fields, magnetic fields, electromagnetic fields, gluonic fields, nuclear weak force fields or strong force fields (nuclear glue), or gravitational fields (cosmic glue). All these fields can exert a force.

**** Not much is known about the interaction of gravitational forces on and within submicroscopic particles, It has been speculated that they may not be affected by gravity.

Explaining Matter as Waves

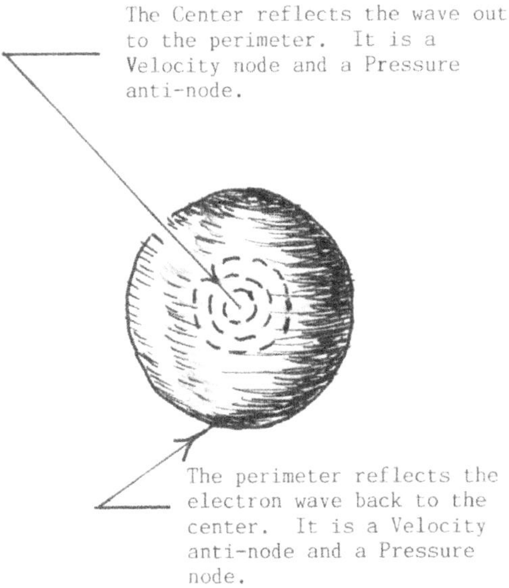

FIG.__10__ The Spherical Electron "Matter Wave" a Standing Wave Structure

In this book, **"Matter"**, is defined as a **"standing wave structure"**, in primordial substance. A **"matter wave"**, I believe is a resonating spherical torsion wave, such as that of the electron illustrated above. As a reminder, **a node** in a standing wave **is a place of calm**, a neutral zone of the type stated. Example a "velocity node" is a place of little change in velocity, a pressure node is a place of little change in pressure. **An anti-node** is the opposite, **it is a place of change**. A "velocity anti-node" is a place of greatly changing velocity, a "pressure anti-node" is a place of greatly changing pressure.

When light encounters **"Matter"** it can be absorbed, emitted, reflected, diffracted, or refracted. When traveling electromagnetic waves meet other traveling waves they just pass through one another. Some believe the forces of nature are carried by "Wave Fields", fields of wave energy, and that particles such as the electron, are packets of wave energy, therefore **matter, is made of resonant wave clouds** in the primordial substance of the cosmic ocean. In other words, Matter is the primordial substance of the cosmic ocean, undergoing a process of localized standing wave activity much as **standing waves** reside in the tubular barrel of a musical horn when played, or along the tensioned string of a played musical stringed instrument, such as a sitar, balalaika, or violin. Light waves are **traveling waves** in that same primordial substance. Light waves can travel to the limits of the universe, unless absorbed by matter (an electron here is considered "matter". If there were no waves at all in the substance of the cosmic ocean, no "thing" would exist in it, only quiescence, unchanging, frozen space and time. The Cosmic Ocean's primordial substance, is infused with energy, **"restless energy" is ever present,** driving substance into activity, some chaotic, some orderly wave activity, and amidst all that activity, "things" like matter and radiation appear when chance is favorable to their forming.

Explaining Matter in terms of Particles

The so called "standard model particle theory" speculates, matter is made of twelve basic particles of which **the electron** is one.*
Those that explain matter in terms of "particles" consider Space as an empty volume, an empty container, except for any particles it may contain. The particle theory people found they needed **a something** to carry the forces of nature, since they had no substance between particles to carry the forces, so they invented **massless particles** (particles without mass.) to carry those forces. Force conduction in the particle theory is accomplished by particles bumping into one another like the balls in a billiard game, their rate of motion (kinetic energy) adding or subtracting, their potential energy, is in their particle mass (their energy level) which can combine or break apart (divide). The perimeter cushions in a billiard game act as force reflectors, much as a mirror does for light waves.

Note: a billiard table has no pockets, a pool table has pockets, otherwise their tops are similar, the best players in both games play, to position the ball(s) for the next shot, not just for the present shot.

The five Massless "Energy carrying Particles":

The Photon - To carry electromagnetic forces, a light quantum ,a name for a discrete unit of radiant energy, an electromagnetic wave packet. A photon is said to have zero mass

The Gluon - To carry the nuclear strong force also termed the strong inter-quark force.
The W and Z particles - To carry the nuclear weak force.
The Graviton - To carry the force of gravity.

And yet one asks, **Do these invented massless particles have a volume, if so what fills it? If these particles have no volume, how can a volumeless particle exist?** Maybe only in a math book.
These energy transferring and holding particles themselves may be an energy wave field. The entire "Mass" of the universe, the mass of it's "Matter" some speculate, is in it's energy, in it's "Fields of Force", in the motion of it's primordial substance.
By one estimate 95 percent of the mass of a Proton is in the proton's **"Gluonic Field"**, the field that holds it's waves together.
It has been speculated by Gabriel Lafreniere in his work "Matter is made of waves" matter is a form of wave activity. Mr Lafreniere's work has been published on the internet If I understand Mr. LaFreniere's theory correctly the following appears to be described in his work:

> **There are particles, or resonating wave structures, with Mass**

The Electron is a resonating wave structure, I myself believe it is a resonating torsion wave.
The Positron is visualized by some as an electron wave structure that is 180 degrees out of phase with that of the electron.

The Proton consists of six equidistant electrons surrounding a centered positron, all held together by, gluonic field forces (gluonic field energy). One can visualize the Proton as a cube with an electron centered in each face and a positron in the cube's center.

The Neutron consists of six electrons positioned equidistant from one another. One could visualize The Neutron as a cube with an electron centered on each face, and no particle in the center. The difference in mass of individual electrons and their combined mass in a proton, or neutron, is ascribed to the presence of the interacting gluonic fields. The mass of a proton, is 1,836 times greater than it's electron components. It is not strange at all that these resonating fields have mass, apparently reinforcing one another, after all the electron itself is a resonating field.

* The book "The Search For Infinity " By Gordon Fraser, Egil Lillestol, and Inge Sellevag. Published by Facts on File, copyright 1995. gives the names of the twelve basic particles and historic milestones in physics and astronomy.

Between the wave people and the particle people there was a need to explain and map **"The Transmission of Force"**. Substance is needed **to transmit force**, a push or pull. Substance is needed **to hold a field of force.** Substance is needed **to transmit energy and information.**

I believe, a **Field of force, is the Strain held within substance that has been deformed.** A field of force in substance holds **potential energy**.

kinetic energy, is substance in motion, this can exert stressing forces.

"The Field of force" Concept

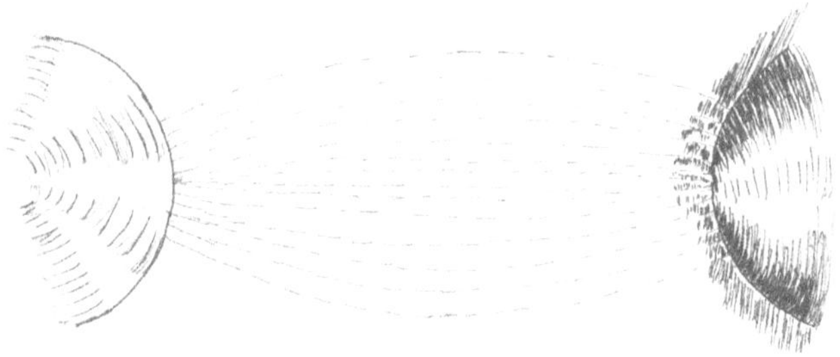

FIG._11_ The Field

One pole of the field is an excess of something, **the opposite pole** of the field is a lack of the same something. Between these two complementary extremes is a "field of tension in the primordial substance".

a) **"The Field",** concept can be thought of as **a property of space,** a state of tension in an **elastic substance,** or a property of a particular volume of space. A "Field", can separately and simultaneously transmit multiple "Field", forces without affecting one another. When those forces are no longer simultaneously present in the same volume of space, it's substance reverts to it's former condition of displacement. This is attested to, by light waves from many sources traveling across the vastness of the cosmic ocean, on reaching earth their places of origin, can be clearly seen, the stars. If they had blended together on their travels, we on earth would only see an illuminated fog.

b) **"The Field",** can be thought of as potential counterforce, a latent force. The Field is a holder of Potential Energy.

c) **"The Field",** can act as a glue to hold (pull together) particles of complementary opposite sign, and repel (push apart) particles of like complementary sign. A **magnetic field** has the potential to react to a change in it's velocity, An electric field is actualized effort to bring change. The Field or the space of "The Field" is the transmitter of the Field's forces. **Force**, is not only a push or pull, it is a transmitter of information.

Nuclear strong field of force, **"The Nuclear Glue" "the Gluon field"** or gluonic field holds together that which makes-up protons, and holds together an atom's nucleus. This Field of force is 10 to the 41 power stronger than the gravitational force (at the very short distances involved in the atomic nucleus).

The weak force - is associated with neutrino interactions.

Electromagnetic Force an **Electromagnetic Interaction Force,** a **Binding Energy** that holds electrons near the nucleus, attracts electrons to protons repels electrons from electrons, and repels protons from protons.

I believe in the end, the above atomic particle fields of force are due to the interactions of "Matter Wave Packets" with one another.

The present speculations of physicists on the nature of matter are all well and good, however. as anyone knows that has physically broken a rock, it always seems possible to break and grind the rock into ever smaller particles without discovering the overall nature of the substance of the rock. Much as scientists are now "smashing atoms" and colliding particles, to end up with ever smaller particles and perhaps fragments of particles without learning their true nature, their true composition, that of which they are made, and that which animates them. Those that are exploring the wave theory of matter have the problem of no workable theory to explain just what is "Waving", is it space itself? The particle theory people had to dream-up fictitious particles to carry energy from one particle to another.

I believe the process of undulating wave motion, occurs in the substance of the cosmos, in the substance of space and time. Energy's power, drives the motions of the waves. Radiations are "packets (bunches) of these waves. Matter such as electrons and protons and neutrons are resonating wave motions in volumes of space, much as sound resonates in a Helmholtz resonating chamber.
In the undulating wave theory, it is thought half the energy is in the form of potential energy (in a displacement strain field), and the other half is in the form of motion, kinetic energy (in a stress field).

The theory put forward in this and my previous two books gives a basis for scientists both armature and professional to work from.

But this sidesteps the really interesting question
Which is, "Of what, do light, or photons, or quanta consist"?

Since energy is motion, what is it that is moving as waves? Whatever it is, it can move in the vacuum of outer space where there appears to be nothing for wave motion to take place in. To answer this question, scientists proposed a continuous theoretical substance "The Aether", an elastic solid, a "medium", that let planets pass through it unhindered, yet was as ridged as steel in order to carry electromagnetic wave motion. No one to date has discovered the Aether. Today some scientists have suggested, it is **space itself that is the substance,** which has the needed elastic properties. It was there in front of them all the time, they just did not see it. The concept of "The Field" is useful. to carry undulating wave motions of radiations such as those of the sun's heat, and light, and other forces, across apparently empty space,
Another group of scientists objected, they said "no, no there is no Aether and you have not proven it exists, so there!". This left nothing to transmit light and heat across the vast distances to the stars, so these other scientists invented little servants, little massless imaginary particles to carry heat and light and other forces across empty space. These imagined particles are given suitable properties to explain observed fact, so there! Speaking for myself, I believe force carrying particles are simply a

convenience to mathematically handle packets of energy with specific energy values., and are no more real than the concept of the Aether. I rather liked the Aether, it at least was continuous and filled all space, little imaginary force carrying particles did not explain what the space between them was, did not explain a continuous "force field", a gravitational field, magnetic field, electric field, or a gluonic field, these "fields", fill the space they exist in. The theory in this book sees space as the continuous elastic substance that carries wave motion across expanses of space, across volumes of space.

** If you care to find further reading on "The Aether", you might consult two rather informative works I rather like, from the golden age of Ether investigation by English physicist **Sir Oliver Lodge**, **"The Ether of Space"**, copyright 1909 originally published by Harper & Brothers New York and London, now reprinted by Kissinger Publishing. And a second book by sir Oliver Lodge, **"Either and Reality"**, preface by Sir Oliver Lodge dated 1925, originally published by George H. Doran Company New York, now reprinted by Kissinger Publishing. A publication of interest is **"A History of the Theories of The Aether and Electricity"**, by **Sir Edmund Whittaker**. Published by Thomas Nelson and Sons Ltd. Copyright 1910. This is a two volume set.
Even **Professor D. Mendeleeff**, of chemical element periodic table fame, had a go at, the Ether with his **"An attempt at a Chemical Conception of the Ether"**, c 1904. Also see a more recent work by Milo Wolff, **"Exploring the physics of the unknown**

Universe". Published by Technotran Press, copyright 1990.

A fellow that could not get rid of the Ether, because reality is a stubborn thing, was **A. Einstein**.
In the theory in this book I have not included the name **"Ether", or "Aether"**, due to all the old "baggage" of argument some have attached to that name, instead I have given the theoretical Ether's elastic properties to Space and Time, which is animated (or if you prefer, "manipulated"),by the aspects of energy, conveyed within the substance of space and time, by the forces of nature.

The reality of space, time, and energy, are how what is before us is explained. Space and time are invisible to the eye, If you had a bag full of each, what would you have? Therein lies the problem of saying what they are. Proving there is an Ether is like that, all who have tried, and some were very capable persons, all failed. There is no dishonor in that, I thank them for trying. There are after all, some things in this world man may never completely understand, or know (experience) directly.

Inertia and Mass

A fair amount has been said by others about **energy** and **mass**. The science of the eighteen hundreds labeled the **resistance of matter to being accelerated, "inertia",** or put another way, Inertia is the difficulty to increase, or decrease the velocity of a lump of matter. Objects of greater or lesser weight, had greater or lesser amounts of inertia. No

cause for that type resistance was apparent, it was not due to friction, as the "inertia, of an object was the same in a vacuum or in an atmosphere.
Other names for "Inertia", are: "Kinetic Reaction Force", "Inertia Force", and "Inertia Reaction Force". Isaac Newton to his credit said he did not know the cause of inertia, but proposed "Inertia", was **due to an "Innate Force"**, and left it to others to specify the source of that force.

Today physicists say, Mass is a measure of the "Inertia", of a body. Still no explanation for this "inertia", property of mass is given.
All Chemical Elements are "Matter", and all possess "Mass", as defined "Mass", is a quantity of matter cohering together so as to form one body. Others have stated "Mass", as the property of a body that measures it's ability to resist any change in it's state of rest or uniform motion in a straight line.
A body or mass of matter that is rotated about a center with uniform circular motion, is said to be "continuously accelerated". In some physics text books mass and inertia are synonymous terms.
In the theory in this book, resonating wave structures, or "matter waves", such as the electron, proton, and neutron all possess "Mass", and have "Inertia". **Inertia itself in the theory in this book is due to acceleration or deceleration of matter. Put in other words, Inertia is due to the displacing and undisplacing of primordial substance.**

The inertia of society.

Inertia also exists in the field of behavior as "the mores", of society. There is a resistance to a change in accepted behavior. Newer behaviors, apply stress forces to society, in their effort to displace accepted behavior, causing strain forces in society, that resist the newer behaviors.

In the theory in this book, **Inertia is due to the amplitude of displacement of primordial substance**, and the **Strain level due to that displacement.**

A mass has it's own internal strains. The space, including the space of a vacuum, has a displacement. The set of displacements called matter when passing through a volume of already displaced primordial substance meets with a resistance to it's propagation rate, it is like plowing through a substance with viscosity, rather than a substance with no viscosity.

Energy and Resonance

"Resonance" is a condition in which energy alternates it's two aspects, between it's kinetic energy and it's potential energy. When substance is in resonance it alternates between being displaced and being un-displaced, alternating between a stressed condition and a strained condition. An electron, is an example of a resonant standing spherical wave. The electron's alternation is so rapid, it appears as a blur (note in a photo of an electron wave a misty look). The electron is a very stable form, it is a resonant wave with almost **no "losses"**, once the electron's wave is taking place, it continues to resonate perpetually, until outside forces stop it, such as colliding with a positron. An example of wave motion with almost no loss is that of electro-magnetic waves, as they travel from near stars, to the far reaches of the universe only absorbed here and there by the small amount of matter along the way, the main reason they diminish in intensity on leaving their source, is their passing through ever greater volumes of space as they leave their sources. Their intensity diminishment is calculate-able using the inverse square rule.

Question: Is there substance that is not in motion in the cosmic ocean? As a practical matter "No". However the least motion in a displacement wave occurs momentarily at, maximum displacement, it's greatest velocity occurs momentarily at, minimum displacement.

Comments on the nature of Time

There have been some who have speculated that Time itself can go faster or slower depending on the motion of the observer, and that space can curve, some say this speculation has been proven by experiment. You of coarse will believe what you wish. For my part I believe their "proof", may be due to misinterpretation of the situation, or experimental error.

As we observe wave motion, light waves, sound waves, or water waves, their propagation is not instantaneous, it takes time. **The Doppler Effect is a result of motion.** * The Doppler Effect is reproducible and has been proven repeatedly with different types of waves, light waves, sound waves, water waves. It does not depend on the expansion or compression of space itself, or the speeding-up, or slowing-down of time. The essence of the

Doppler effect is, the difference in the number of wave pulses per second emitted by the source, and the number received per second by a listener, **as the source and listener** are moving closer together or farther apart.
We experience this effect by the rise and then lowering of the pitch of a whistle or horn as it's source speeds past us. Similarly if waves are traveling across a lake at the rate of 'X' wave fronts per minute, a boat running against the travel of wave fronts, or running with the traveling wave fronts, will experiences more or less than 'X wave fronts per minute slapping against it's bow, in fact if the boat is moving at the same rate as the waves, it will experience no wave fronts at all hitting it's bow.
Someone standing on the shore admiring the waves, would say there rate of wave travel across the lake surface was constant.

* Doppler Effect - discovered in 1842
by Austrian physicist
Christian J. Doppler.

It has been speculated that a resonating spherical electron wave moving beyond what acts as it's resonating chamber, will act much as the sound from a musical horn moving out from the bell of the horn to a listener. Also, that the three characteristics of the spherical electron wave field, when moved bodily through space, will suffer a distortion due to an internal Doppler effect, particularly when moving near light speed.

Picture the resonating wave moved bodily through space, The wave "feels" the conditions in the space it is moving through. The electron resonance, will compress ahead of the center of resonance as the wave moves forward, and expand on the trailing side of the center of resonance. Overall the three dimensional spherical wave would distort, and assume an egg shape, with it's long axis, in the direction of bodily motion, A resonating egg, the faster it's motion through space the more elongated the egg becomes. I would emphasize that space or time are not compressing or expanding, only shape of the wave field with it's three characteristics changes in the space they occupy.

The Space and Time of this Theory

Space and Time, the two aspects of primordial substance. Space and Time in this theory, each have two halves to their whole, that would be four parts in all, two moveable, two not moveable. But and I emphasize this, both together, Space and Time, make up "A single **Substance**" **Primordial Substance**, Primordial Substance is the "substance" part of the Cosmic Ocean. To picture this situation as an analogy, look at this cosmic ocean as if it were a coin, one side moveable the other not. When energy causes a torsion displacement, as the moveable half moves out of it's original relationship to the non-moveable half The displacement causes "strain forces", build in the substance of the coin. When the displacing "stressing forces", are removed our coin's halves spring back into their original alignment.

When energy causes displacement of space "strain forces", appear, we interpret these as "forms", and the clock of time starts ticking for the displaced space substance, it's displacement has a limited duration. The actual duration in seconds or years depends on the volume affected and the amplitude of the displacement (the degrees of torsion motion).

Matter and the Science of Cold

One of the great challenges of science was to stop all thermal motion of atoms and molecules. The way to do this was to remove all heat energy that was causing the invisible jiggling around of the particles, this stopping of molecular motion is believed to occur **at minus 273 degrees Centigrade** (that would be **at a minus 459.67 degrees on the Fahrenheit** scale, **or at zero degrees on the Kelvin scale,** zero Kelvin degrees is also termed **"absolute zero"**.

Pardon I am showing my age, the Centigrade scale has now been renamed the Celsius scale. Same numbers, just a name change. Scientists have now come very close to "absolute zero".

At room temperature, and a pressure of one atmosphere, the element Chlorine (Cl, #17) would be a greenish-yellow gas, at minus 34 degrees centigrade, it would be a yellow liquid with a slight greenish tinge, and at minus 101 degrees Centigrade, the Chlorine would become frozen solid. Asking, what would happen if we lowered the temperature farther, and farther? If temperature is motion, what motion remains after the dancing around of molecules, atoms, and electrons is removed? Visualize the **"matter wave"** motion of a Chlorine electron, even though the electron is not moved from place to place. The matter wave of an electron, is the wave motion of space displacement itself. If wave displacement motion would cease, the displacement would as it were be frozen, "it would be an unmoving wave, a wave frozen in place." Ah! We, are at the nub of the matter now. It seems at this point all energy would have been extracted from the Chlorine matter waves, all that remains would be, is an unmoving volume of **"Primordial Substance".** "kinetic energy would have been totally removed, all that would be left would be the frozen potential energy "strain forces" held in displaced substance, the chlorine atom waves would cease to wave, it would be a standing wave without it's velocity component. I leave to you to say what would happen to what remains, in any case, **"Chlorine",** would no longer exist. It would be like an illusionist's act. The thing is, I doubt all energy can be separated from pure primordial substance, it would be like removing Yin from Yang,, or removing space from it's relationship with time, or saying one side of a three dimensional object is still possible without it's opposite side. It is not logical.

What I do suggest can occur, is that what appears to be energy loss, is really a shift of the proportion of kinetic to potential energy in the space that was the chlorine atom. This suggestion at present is difficult to verify as the "amplitude or magnitude" that space is **"Displaced",** presently it is not directly measurable, there is no device to do so. The space, in the freezing chamber might well appear to be empty.

For now, quantum physicists believe "**atomic particles such as a photons or electrons have no quiescent state,** they are always tirelessly moving, forever moving **if free**." They add that if their motion is influenced by other particles, or force fields, then their motion, is no longer free, it is then "**restricted motion**".

The motion energy of freely moving particles, such as photons and electrons is termed, **"zero point energy", freely moving particles are immersed in a sea of particles in a similar situation,** much as a deep sea diver is immersed in the press of water all about him. Waves (like electromagnetic waves) that are freely radiating, freely propagating, through space crisscrossing one another, without affecting one another would also fit the definition of unrestricted motion and possess zero point energy. As I understand a quantum physicist's view of zero point energy, it does not include, the energy **within** the "**matter wave", of a particle**. which is wave motion that is reflected back on itself, a resonating wave, endlessly moving, endlessly reflecting, endlessly engaged in wave motion.

Using Zero Point Energy

There have been attempts to use zero point energy, Some have compared these attempts to trying to use of the pressure at the bottom of the ocean, while at the bottom, or trying to use the flow of a moving river while you are moving with it, or using the value of a gold bar confined in a safe. Gold that is confined possesses unused value.

It should be observed however that flowing water, moves pieces of matter caught-up in it's flow, and that water under pressure, has the effect of depressing it's boiling point, that has shown usefulness.

Some day I believe man will control the energy held **within** resonating wave bundles, within "matter waves", man will then have the ability to reform these packages of "held" energy, to accomplish man's needs and desires. Perhaps extract the radioactivity from a substance and render it no longer radioactive. One way might be to alter the resonant frequency of the radiating atoms. It's a thought anyway.

When Chemistry was still Alchemy, **Stahl** and **Becher** in Germany, about 1690, came up with an incorrect but imaginative, and at the time seriously believed by some, **Phlogiston theory**, wherein when a substance was burned, the fire principle **phlogiston** escaped, **and** only **ash (calx)** remained of the burned substance. If only a small amount of ash was left, most of the substance must have been, Phlogiston. If a lot of ash remained after burning, little of the substance must have been Phlogiston. At the time, to most persons, air was air and took no part in the burning process. A few Chinese about 400 or 300 B.C. suspected two kinds of air existed, an active type, **Yin air**, and a passive type, **Yang air**. **Leonardo da Vinci** (1452-1519 A.D. Florentine artist and scientist) wrote of active and passive air. We now know that **99 percent** of the Earth's "Air" volume consists of four components, two of which are very active gasses, animals .breath **oxygen** to live, and plants, breath **carbon dioxide** to live. In a fire, oxygen combines with parts of the substance burned to produce various oxides including carbon dioxide, (C O2, one atom carbon attached to two atoms of oxygen).

Components in "Air" consists mostly of Water Vapor and ten gases

Nitrogen (about **78 percent** of a volume of air))
Oxygen (about **20 percent** of a volume of air)
Carbon Dioxide (amount in atmosphere air varies)
Water Vapor (amount held in "air" varies with temperature and pressure)

Only one part, out of every one hundred parts of atmospheric "air" is a mix of the following seven gases.

Argon
Neon
Krypton **
Helium - Helium was the most difficult of these gases to solidify. Solidification was achieved directly from it's gas, at a temperature of minis 268.9 degrees Celsius, and a pressure of 140 atmosphere. In the atmosphere Helium is rare, out of 10,000 volumes of air, only four volumes would be pure Helium. Commercial supplies of Helium are from natural gas wells.
Xenon (pronounced Zee-non).
Hydrogen - there is very little hydrogen gas in the atmosphere, most is in water and mineral compounds. Water is made of two parts hydrogen and one part oxygen. Oddly hydrogen gas is insoluble in water.
Radon a radioactive, colorless, odorless, gas, only a trace is in the atmosphere.

Legions of the Air

Eleven legions traverse Earth and Sky
Delivering Nature's messages
On northern nights their colors wave
White and blue, green, red, and grey. ***

*** The colors referred to, are those generated in an oval shaped zone **43 to 200** miles (69-322km.) above the earths magnetic poles, these
Auroral lights, appear as beautiful curtains of light gently waving in a breeze of streaming protons and electrons from our sun, these excite the gases of earth's thin high atmosphere, the excited gases emit the various colors. Nitrogen's blue, purple and red light, Oxygen's yellow-green or red light. White light results from a mix of colors, grey is a dim white against a black background. If you wish to see this wonder of nature, and want a few pointers on where and how to view it, ask an internet search engine, or speak with a lecturer at a Planetarium, or with an astronomer.
A warm jacket and cap, or an insulated "Hood-dee", garment, may be good to have along, viewing on clear nights, tends to be cool, especially north of the 45th parallel

** The chemical element **Krypton,** atomic number 36, is a colorless, odorless gas, **not the fictional green glowing solid Kryptonite** of Superman comic stories.

Note: The Roman Army had legions. Augustus Caesar reduced the number of legions from sixty to twenty eight. In the 3rd century A.D. the legion size was reduced to 2,000 men. The cost of supporting a military, of about 500,000 men, lead to Rome's economic downfall.

Superconductivity - Is the is complete disappearance of normal electrical resistivity to electric current flow in metals, alloys, and compounds, when their temperature is lowered to near absolute zero, discovered in 1911 by H. Kamerlingh Onnes. Each metal or alloy has it's own temperature at which it becomes superconductive. Superconductivity occurs in the element Lead at 7.26 degrees Kelvin. A Niobium (#41)-Gemanium (#32) Aluminium (#13) alloy, has a superconducting temperature as high as 20 degrees Kelvin. Medical Magnetic Resonance Imaging (MRI) machines have a huge current flowing around a superconducting path to make possible the strong magnetic fields needed for imaging. Once the current is set flowing, it will persist as long as the path remains at superconducting temperature.

The Arbitrariness of setting the "zero point"

Many decisions in life depend on which side of an arbitrary zero point a thing falls. We have to start somewhere to compare related things to, or changes that occur. **Arbitrary decisions are based on reliability, personal preference, discretion, or convenience.**

In physics, the point on a temperature scale, designated zero, has an arbitrary basis. To measure temperatures one of three scales are commonly used:

The Centigrade scale is also named **the Celsius scale.**
The Fahrenheit scale, and
The Absolute or **Kelvin scale.**

Absolute Zero, on the Kelvin scale is at, zero degrees Kelvin.

Absolute Zero Kelvin on the Fahrenheit scale is at minus 459.67 degrees Fahrenheit.

Absolute Zero Kelvin on the Celsius-Centigrade scale, is at minus 273 degrees Celsius.

Below are shown the boiling and freezing points of pure water on each of these three temperature scales, at a standard pressure of:
760mm (millimeters of mercury),
14.7 p.s.i. (pounds per square inch).
101.4 kpa (kilopascals, thousands of pascals).

Pure water's Boiling point is	Pure water's Freezing point is
+100 degrees **Centigrade.**	**Zero** degrees Celsius
+212 degrees **Fahrenheit.**	+32 degrees Fahrenheit
+373.15 degrees **Kelvin**	+273 degrees Kelvin

As molecules are cooled they move slower and slower, and also become closer together. At absolute zero molecular motion stops. All known gasses become solids at absolute zero. It has been discovered that light is slowed when passing through transparent material at absolute zero, so much so it's progress can be seen as it moves through the substance. **Absolute Zero** is at the following: temperatures, **Zero** degrees **Kelvin,** equals, **minus 273 degrees Celcius,** equals, **minus 459.4 degrees Fahrenheit.**

The **wave motion** that makes an electron an electron, or a molecule a molecule, **still goes on at absolute zero**. If this wave motion stopped, the molecule or electron would disappear, it would no longer be an electron.
I believe, there is no place in primordial substance where one of energy's two forms, or aspects , kinetic or potential energy does not exist. Put another way, if there is substance there is energy also.
In mechanics - The zero point of a swinging pendulum's motion is arbitrarily set at it's point of maximum velocity, the mid point, the bottom of the swing, not at zero velocity, at the tops of the swing, which would be at the beginning and end of the swing.

In electronics, which is concerned with current flow, the current flows from the power source positive pole, the pole with an abundance of electrons, around the electric circuit and into the negative pole, the pole with a deficiency of electrons at the power source. The polarity of these poles in electronics was chosen arbitrarily.
A Question: is there a place of zero current flow in an electrical circuit, or a battery in which current is being drawn?
In the physic's that studies particles, matter, waves, and the forces that animate them, the first particle discovered (the electron), was assigned a quite arbitrary negative designation. **The charge given to all other particles is based on** weather they are attracted to, or repelled from an electron. Opposites attract, likes repel.

As a matter of interest, the electron has the same "sign of charge" as on an electrically charged rubber balloon, both have a negative charge.

Where is the zero point between chaos and cosmos? I would place it where both are minimum, on one side of zero, chaos would increase to infinite diversity, on the other side of zero, Cosmos (order) would increase to a singularity. The cosmic ocean existing as infinite diversity, and the cosmic ocean as a singularity, some may say would occupy an equal volume, though this is no sure thing. If space and time were "melted", together, as would be the case in a cosmos with no dualities within it, a true "singularity", what would the result be like? Would it collapse to zero volume?

An interesting side light of an electrical nature

On vacation in northern Wisconsin our family stayed at a wooden frame cabin by a river, it was a wonder filled existence. One rainy afternoon my father was standing near the kitchen sink observing the scene outside a window above the kitchen sink and hearing the sound of rain on the roof, when there was a brilliant flash of light outside, accompanied by an very impressive boom, lightning had struck a tree stump sixty feet (18.3 m) away. Of a sudden Dad looked very surprised, he had felt pin like pricks on his chest, I was only three feet away at the time and on checking found three or four light brown burn holes in dad's white tee shirt. Apparently the great surge of current in the lightning bolt had grounded into the lawn watering faucet pipe out near the stump, he may have received an induced secondary current from the metal window screen. My dad was fortunate, one can only guess the voltage of the current that burned the holes in his tee shirt, it jumped a gap of three feet through relatively dry air in the cabin. A rough estimate of the voltage would be, **1,000 volts, to arc across a 1/4 inch gap** (0.6m.). To arc across a 3 foot gap would need about:

3 feet x 12 inches x 4 one quarter inches x 1,000 = **144,000 volts.**

An Electric Question

Why are there more negative electrons than positrons roaming about the universe?
The difference between an electron and a positron appears to be one of the "Phase" of it's resonating wave. The thing is the universe as a whole may be engaged in a single great torsion wave motion that has a direction to it's twist, a direction to the twist of it's displacement motion. It may be this **background twist** that is encouraging survival of more electrons than positrons. This great torsion wave cycle of the universe may take billions of our years to complete. **a simple way is needed to measure the amount space is displaced, and the direction of the displacement**, it's **direction of torque, it's twist (handedness of rotation), and it's velocity of motion.**

Future possibilities

a) - Make an artificial electron, either of the usual size or a larger version.

b) - Make Matter directly from "primordial substance", by using energy. This would be like making matter out of what is now considered "Empty space". Mining and despoiling of the land would no longer be necessary to sustain life. A source of oxygen and food for space travelers could then be made on route.

c) - That messages will one day will be sent at faster than light speed by artificially increasing the strain in a beam like column in the cosmic ocean's primordial substance. Conversely light at slower than what is now considered light speed will be possible. This by reducing the strain in the primordial substance, the light is passing through.

d) - The ability to slowly drain the internal energy of an electron or atom using it as an energy source like a super battery. **Some caution,** and provisions will probably be needed to prevent excessive charge build-up in the energy acquisition zone, due to depletion of electrons, or phase change of electron matter-waves.

If one could latch onto the power of eternal change in the cosmos, the energy that can displace primordial substance, the energy that powers the movement of chaos and all else in the Cosmic Ocean, then that person would have the power, to cool or heat homes or propel man to the stars, without bringing ruin to our planet by excavating it for minerals or polluting it with waste products.

If you believe "everything has been found out" about the natural world we live and work in, consider what authoress, researcher **Jenny Randles** has termed **"Time Storms"** in her book **"Breaking the Time Barrier."** page 260. Copyright 2005, ISBN

13-978-0-7434-9259-1, I believe she is a very careful researcher, doing her best to determine the facts of any occurrences given in her books, you may even find the reference section near the end of the book interesting.

We have now spoken of the great cosmic ocean with it's **three basic components**, two of which are, a duality consisting of a complementary opposite pair, **substance** and **energy**, and a mysterious third component, **the messenger** that links and facilitates communication between the other two. Figure #12 A-B is a Matrix diagram of the Cosmic Ocean.

Notes

Matrix Diagram -A –The Cosmic Ocean (Universe)

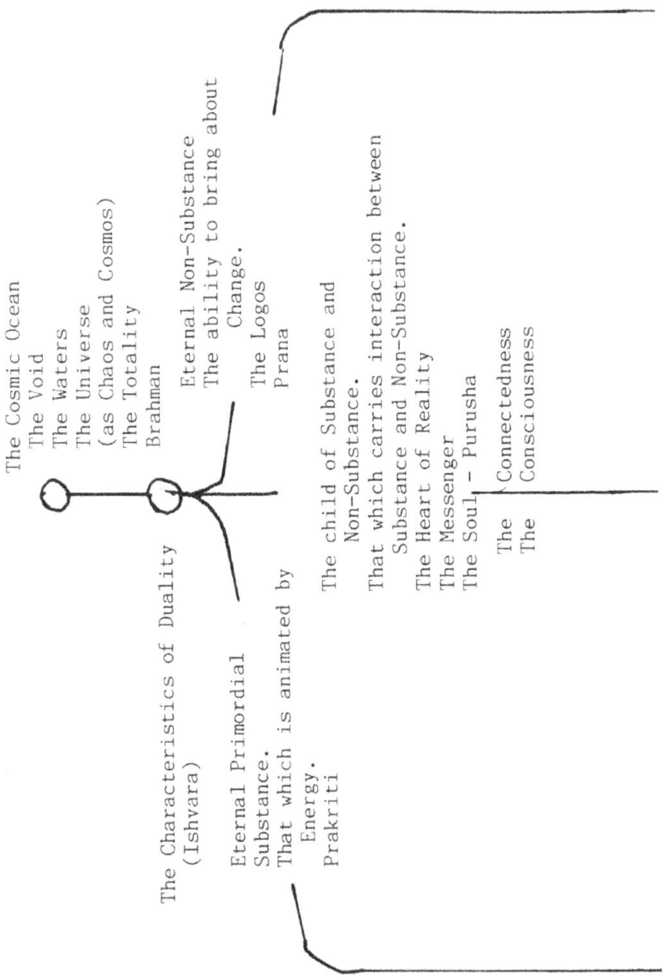

FIG. 12 A Matrix Diagram of The Cosmic Ocean

Matrix Diagram -B The Cosmic Ocean The Universe

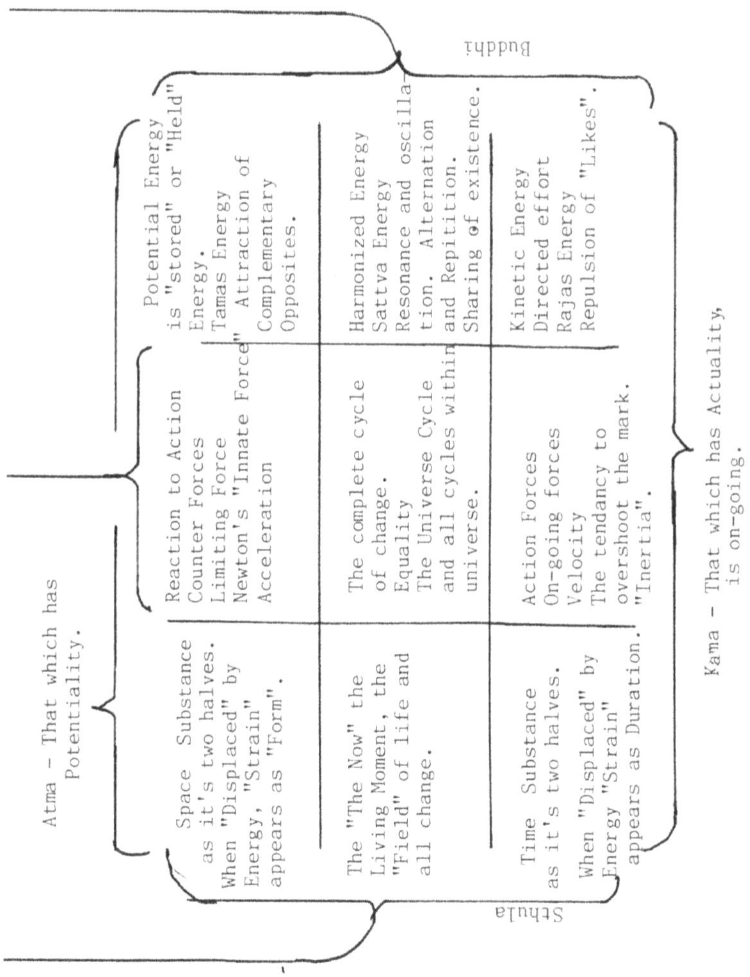

FIG. 12 B Matrix Diagram of The Cosmic Ocean

Man, a small thing in a vast Cosmic Ocean

Now I will mention **a relatively small thing compared to the cosmic ocean, the universe, that thing is "Man"**, puking,** mewling,*** demanding, Man, that conglomeration* of parts, that can act as a unity, sense the situation within and outside itself, and say:
"I exist, I Am, along with that which is outside of, or "other" than myself, and yet I am part of it".

* **conglomeration** - a collection. A mixed coherent mass.
** **puking** - vomiting, throwing-up the contents of one's stomach.
*** **mewling** - making whimpering or whining sounds.

Note: The "**mores**" of society. Mores are the fixed customs or folkways, habits, or manners that a particular society or community expects an individual to conform to. The conditions under which a society or community lives may change over time, "mores" may also be changed to adapt to changing needs or opinions.

Does Man Somehow Mirror The Cosmic Ocean?

Before looking more closely into the nature of man, I believe it will be helpful if we compare the Cosmic Ocean (Universe), and Man.

1) Cosmic Ocean has three basic components, and so does the human Self.

Cosmic	**Ocean Man**
Primordial Substance	Physical Body
Non-Substance (Energy)	The Spirit, The Will
The Connectedness	The Soul

Man's similarities to the universe:

a) - Man is composed out of the **same "stuff"** as the universe. (Substance and animating Energy)

b) - Like the universe man is a **self regulating mechanism** (internal temperature control, fluid flow etc.).

c) - Man is **subject to cycles of change**, and so is the universe.

Man's unique qualities

a) - Man has a physically separate, **complementary opposite gender Self.** All duality is within the Cosmic Ocean, it's selves are both within The Cosmic Ocean, it has no outside, no existence, beyond it's perimeter.

b) - Man's body, **senses the world outside itself and within itself**. The cosmic ocean has no-thing to sense outside itself.

c) - Man's Self has **a life span** beyond which, he is no longer whole. **The Cosmic Ocean is "Eternal"** it's life span is unending, it's "Self" is always whole.

d) - Man has many needs to satisfy. The Cosmic Ocean has only one **"Dynamic Equality"**, all change is due to satisfying that one need.

Notes

Part Three

The Self

The Concept of "The Self", The Atman

Body, mind, and Spirit working together, are "The Self".

"The Self" was believed long ago to consist of both Body, Mind, and Spirit, working together. The matrix diagram of "The Self", figure 13 A-B, pages 100-101 ,

The Body, the substance of "The Self", is the physical vehicle that transports the forces of nature.
The Spirit, energy, is the animator of the substance of the Self.
The Mind is seen as the interpreter, regulator, the controller.

The body's sensations apply forces to modify behavior of the Self.
The mind interprets these bodily sensations as pain and pleasure, as "Needs" or "Satisfactions", as fears and desires, The Mind directs the body to act to it's advantage. To seek safety, avoid danger, appreciate pleasure and to avoid and fear pain.

Man is a civic animal, and needs to not only be useful to himself, but to others.
Since man does not exist alone in the world but with others of his kind, **the forces** that control decisions in the mind, **are a person's needs, and expectations,** also being a civic animal, **consideration of the needs and expectations of others** so that all may live in harmony. A certain amount of strife, and need will always exist, the thing is to distribute it with equality as much as is practical.

Most persons wish to live in harmony. They strive for a dynamic balance between the urgency of a need, and satisfying that need, the mind sets priorities to different needs, applying greatest efforts to the most urgent need or expectation.

History however has shown some individuals, some governments, some religions, and those that run them, **do not always have the good of the individual at heart**, but only their own good, these type persons and organizations **ask for more than their fair share, they ask for that which is beyond their right to ask.** They push aside the common sense wisdom of "Because you can do a thing, does not necessarily make it good or right to do",

Comments on FIG. 13 A-B The Self.

The same groups of squares, of the matrix diagram that represent the nature of, The Cosmic Ocean, as "**Worlds**" or "**Planes**" of existence, are used to represent **"The Self"** not only of the universe, but apply also to an individual person, (a man). Different terminology may be used in the matrix diagram of the "Universe", and that of "The Self", However their interaction of terms is similar, and the same three basic components (Substance, Energy and the link between them Force) are present in both matrix diagrams.

In the matrix diagram of **"the Self"** Figure 13 A-B , The **top horizontal row** of three squares are squares represents the **"Potentiality"** of existence, the plane **"Atma".**

The **bottom horizontal row** of three squares represents the **"Actuality"** of existence, the plane of **"Kama".**

The **right hand vertical column** of three squares represents **"Buddhi"** awakening of "The Self". Buddhi is energy and it's regulation. The regulated animation of "The Self", control of behavior, of bodily systems, heart rate, breathing etc, and the saying of yes or no to action as needed. This can be said to be **"The Likeness"** of the self, where potential abilities meet on-going efforts, on-going Actualities and their resolution takes place.

Left hand vertical column of three squares, represents the plane of **"Sthula"**. Sthula is where the form of the self appears, and the behavior of the self appears. Sthula is **"The Image"** of the self.

> **Caution: Here**, The World of **Atma is not The Atman** (The Self). The Worlds listed below are represented in the matrix diagrams each "World is allotted three squares, the corner squares are shared between two worlds. **Paran-atma is all potential existences,** and all potential worlds, or planes of existence.

Diagrams A and B, make up a complete diagram.

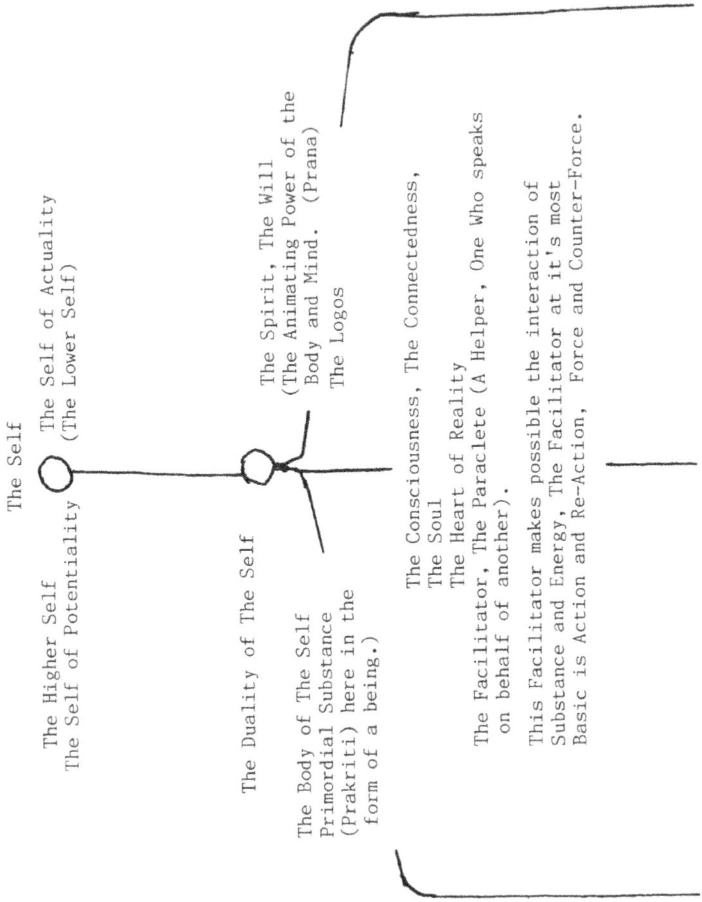

FIG. 13 A . Matrix Diagram of "The Self"

Atma of "The Self" as Potentiality. Re-Action (Feminine in nature.)

Atma also "Higher Mind", the Reasoning, understanding mind. Atma as "The Higher Self".

Sthula / Buddhi

Memory, Storhouse of Experience. Imagination, assembler of sensations. Impressions of things.	The Divine Soul The Subconscious Inhibitions and Mores. Satisfaction and Pleasure. Understanding as the definitions of things.	Phanes - as The Light of Reason. Reason, the ordering of diverse facts. Love as "The Need to be Whole." Making "order" by matching together pairs.*
Truth - as a thing being Experienced in the "Now". The Persona. Behavior	The Human Soul The Waking Conscious The Complete Cycle of Life. Individuated Person	Metis - as Wisdom, Sophia. Balancing of Values. Love as "The Need to Share Existence with one's Complementary Opposite". Harmony - Sattva
Opportunity. Sensations (Sight, Sound, Hearing, Taste, Touch, Temperature, Smell.)	Animal Soul The Id. Automatic body functions such as ... Heart rate, breathing rate, Fear or Flight. Needs and Desires	Ericapaeus - action as Discrimination, Selection, Choice. Judgement. Love as "The need to be "Individual".

Kama The Actuality of The Self. Self as on-going active change.

Kama as Lower Mind and The Lower Self.

* pairs of complementary opposites, matched together to make-sense of diverse facts and enable one to see relationships.

FIG. 13 B . Matrix Diagram of "The Self"

Mind, Manas

Birthplace of the Goddesses and Gods

Mind is the place of regulation of "The Self"

"The Mind", is the director of the body, and it's regulator.
But how humans visualize the parts of a working mind is interesting.
Aristotle and **Plato** saw the mind as **that part of the soul that can "think",** that is, the mind can deliberate and decide, reason and choose. "Mind" is the place of Memory, Imagination, Ideas, and the birthplace of goddesses and gods.
Some have said **"The Mind"**, of an individual is their **"Sixth Sense"**, or their **"The Third Eye"**, the ability is to experience sensations and interpret what they mean, perceiving how a person's life relates to things that influence it.
In Hindu thought "Mind", is a world or plane of existence, The world of Manas.

It is in "the mind" that we are happy or sad, feel joy or sorrow, hope or discouragement, acceptance and rejection, believing and not believing, and it is in the Mind that we **believe what we will believe**.

"The Mind", is the messenger of "The Self",
The carrier of thought through
the substance of the body.

The "carrying of messages in the mind is termed "**Consciousness**", the two aspects of energy flowing through the substance of the mind. This movement of thoughts is possible because of the oneness, the **"Connectedness"**, of primordial

substance. The two, **the "Connectedness"**, of substance that carries information, and, the information carried, the thoughts themselves, the **"Consciousness"**, these form **"The Soul"**, the Self of experience, the Self that feels emotion, The Self that may error, The human "Self". The Mind is like a shipyard, building and launching ships, ships of thought, thoughts for "The Self" to use in crossing the sea of experience, the sea of life.
The "Self" is a reaction to the forces it's experiences, the result is **behavior, attitude,** and **emotion**.

Comparing the concepts of Mind, and the Soul, we find they are different views, of the same phenomena.

The Higher Mind which includes the subconscious (higher mind was formerly the Divine Soul) the mind's memory and reasoning ability. The Higher Mind is a reaction to the lower mind's on-going efforts. **The Subconscious is, "The Divine Soul".**

The Subconscious

The Subconscious is, the mind's world of Potentialities. In this "Higher Mind", are found Memory and Imagination, Understanding and Reason, it is a place where want, hunger, and torment, can be put aside and potential satisfaction seen, The subconscious is where we believe what we will believe, it is a place of satisfaction, of goals as we define them. The subconscious is where can be found "The promised land", or the place known to some as Paradise, and to others as Heaven. These places are as real as happiness. Each person sees them how they will, some see their paradise as a lush green desert oasis with it's pool of life sustaining water, for others it is a virgin forest of soft mosses, ferns , running brooks and the voices of it's creatures.

Some will say these places are only in one's imagination, but consider, **it is in your mind you are happy or sad, it is in your mind you "believe" or do not believe**, and in your mind that a thing is envisioned, and it can be so in your mind, here and now, you have but to picture it's existence as you wish it to be. No this ability will not solve all your problems, but there is a measure of peace and comfort in experiencing it.

Some find their paradise during daydreaming, for others it is found while sleeping, a few use self hypnosis to enter the world of the subconscious,

The Lower Mind which includes **"The Id"**, Lower mind was once termed **The Animal Soul**. Lower mind's activity is not due to reason, it is the body engaging reality for survival, by realizing feelings, urges, needs, and using one's ability to bring change, and achieve the goal of satisfying those needs, desires, and urges. The lower mind is, actualized directed force, to alter and direct behavior, it is **The Id, "The Animal Soul".**

The Id

The Id is the mind's world of on-going Actuality. The Id is the earthy message sending center. The Id (is a world of sensations, of needs) is the **awareness of the effect of one's "existence"**, The Id has also been likened to a swamp, a morass that can, mire or bog a person down trying to satisfy continual "needs" and "desires" of mind and body.

The Waking Conscious
The Mind Between

Thoughts from the Higher and Lower Mind are resolved in "The Mind Between". The connecting regulating "Link".

The Mind Between is **"The Mind of "Emotions", The Mind where forces of nature are resolved.** The resolution of force and counter force, is represented in the central square of the matrix diagram, Interactions are an awareness of "The Now".
Strivings to achieve satisfaction of a goal, are felt as tensions in the mind, emotions of anguish, compassion, and other feelings, these were once thought to reside in one's heart.
"The Mind Between", **"The Waking Conscious"**, **"The Human Soul"**, is bombarded with sensations of the lower mind, and reasoning from the upper mind, bombarded with randomness from the realm of substance, and logic from the realm of energy, here these tensions are resolved into action or inaction.

In the Waking Consciousness life's experiences are presented, and life's problems, so here hopefully a resolution can be arrived at. These moment by moment resolutions require a good deal of attention, and continual adjustment.
The Navaho Indians of North America see the positive-ness in experiencing life, and walking their path through life, say **beauty before me, beauty behind me, beauty to the right of me, beauty to the left of me, beauty above me, beauty below me, I'm on "The Pollen Path"**, the path of life.

It has been speculated, probably by a mathematician in a coffee house, that if a person had **all** the facts about the world we live in this instant, all present conditions in the cosmic ocean, that such a person could predict the future.
Such an airy fairy speculation, leaves out the teeny weenie fact that an infinite vastness would take infinite capability to deal with the problem. The human mind has limited capability, the human mind is subjected to random influences, and therefore not entirely subject to logic. A random occurrence may unpredictably influence pure logic. The English might express this as "Chance putting it's oar in the water of logic". This has led to unexpected results, sometimes discovery, sometimes unwarranted suspicion, or unwarranted conclusions.

The "Mind", of the Cosmic Ocean

The whole cosmic ocean can be understood as having a mind in that it senses and adjusts, it's reacts to on-going efforts bringing change.

The continuum, the oneness of substance conducts the aspects of energy within substance such that one cloud or volume of space affects adjacent volumes, they in effect "feel", one anthers presence and displacement condition. The strain forces of one volume of space are felt by adjacent volumes. The Hindu **"Net of Indra"**, is a similar concept, a net of gems, where each reflects images of the others, where all things are as they are due to others being as they are. Each thing is affected by surrounding things, as a result waves of behavior roll about within the whole.

Seeing the mind as Thesis, Anti-Thesis, and Synthesis

The **Subconscious,** is "**The thesis**", as potential goals, "held" definitions, or views.
The **Id,** is **"The anti-thesis"**, is a policy challenging "held", views, with strivings to reach a new chosen goal. The resultant **synthesis** is arrived at in **The Waking Conscious**. Synthesis is the action of blending together parts, to form a whole. A blending together of moments of time, into a connected series of moments, that form a river of time in which change flows. Hopefully both the needs of "The Self", and their satisfaction, will be accommodated, in this synthesis.

"Tensions" in our personal lives

"Tensions" are due to complementary opposite forces of nature encountering one another. Tensions exist as Stress forces and Strain forces, their interaction molds and un-molds our "Behavior", molds and un-molds the shape and character of "Primordial Substance". The forces that interact **on and within a person** are, **"Stress Forces",** that exert an effort to bring about change, pressures, to minimize need and desire. The other forces are **"Strain Forces",** these are limiting forces (counter forces, the inhibitions a person has, the behavior that is being pushed to change, to satisfy needs.

Please note: **Sensation is fact, Perception, is an interpretation of fact. Perception, is physical sensations interpreted in light of past experiences,** and the mind's judgment as to their meaning and value. Example: a witness heard a sound, and interpreted it (that is perceived it) as a gun shot. Evidence (facts) and other testimony revealed the sound was a firecracker exploding.

The Mind maintains an awareness (a Consciousness) of existence, an awareness of the world within itself, (the mind makes an effort, to know both sensations from within the body and those, the body receives from the world outside itself, the world that is "other", than "Self". The Mind seeks to understand both worlds. As an aide to understanding, the Mind stores sensations and impressions for recall and comparison, to use as guidance in making it's decisions. Some sensations are welcome, and others not welcome.

"The Mind" possesses three main abilities: **Knowledge,** as memories of experiences. **Thought,** as the processing of sensed facts including, finding their relationships, and their relative importance.

Purpose is determining goals and the taking needed actions to accomplish those goals. The Mind recognizes needs, and desires, then devises ways to satisfy those needs in their order of importance and urgency, for an individual to "Survive"**. Survival** is usually of first importance, a first priority as a goal.
Understanding is accomplished by describing (defining) terms encountered, **Reason** compares terms and matches those that are, complementary opposites into pairs. **Imagination,** is the thinking mind saying "what If?" this or that could be, or take place. The Mind **imagines potential outcomes** to "The Self's", situations and probable outcomes if action is taken, then judges desirability.

To say the mind is a self programming computer, made of living biological substance, somehow leaves out **the emotion of life**, the feelings of joy and terror, attraction and repulsion, suffering, excitement, thrills, fervor, and the longing to experience desire, love, hope, and their complementary opposites, revulsion, hate, and disappointment.

Question: Can a computer desire beauty, feel forgiveness, or know the fragrance of Love?

The word fragrance, generally means a pleasant sometimes sweet scent, the scent of a moist pine forest, various flowers, perfume scents, and to a man, in addition to these, the scent of a woman, all these if pleasant are attractants.
Interestingly, the scents found in many perfumes are artificially made by chemists, their use in perfumes, takes into account their ability to not

produce undesired reactions when applied to humans, also considered is scent's persistence after being applied.

Mind and Nirvana

There is a state of mind where one is at peace within oneself and with the world beyond "The Self". The word that has been used to describe that state is **Nirvana**, a word that is a combination of "Nir", **without** and "Vana", **wind**. A mind that, is not in turmoil, a mind, without wind. A mind that has found a measure of happiness in understanding, found a harmoniousness in life. It is like staring out of a window at some lovely clouds and **having no need for more**, a sort of dream state. The state of Nirvana does not mean the world stops for you, life goes on, some seek nirvana using means such as meditation, which can be calming, for others, the nirvana state, can just come upon them unexpectedly.

"The Self" revisited

The idea of "Self", The entire individual. The self of an individual person, or the Self of the entire cosmic ocean. Hindu Sages termed "The Self", "The Atman". The Atman or Self, consisted of, **The body** (substance), **The Spirit** (the power to animate substance), and **The Forces** that carry the power to animate the substance. These forces have been called **"The Breath of Life"** or **"The Soul"**, of the individual. That which brings the substance to Life. These three things are the essence of each individual self, whether it is a person's self, or the supreme "Self", of the entire cosmic ocean, The Paran Atman.

The Mind's Complementary Opposite "Self"

It is interesting to note that Egyptian folklore at the time of the pharaohs (2970 B.C? to 30 B.C.) told of each person having, a "Twin Self" of the opposite sex **"In another land."** Their twin self also had a soul. You may possibly have an actual twin self somewhere if you believe this, there is however a complementary opposite twin self much closer, there is a twin self in one's mind, **in a way the mind is "another land",** a land of potentiality, this twin self in one's mind is termed by psychologists **"The Shadow Self"**. Looked at this way, we are **people of two lands**, two worlds, the earthly world and a mental world, a world of heavenly potentiality. **These "Twin Selves"** are complementary opposites selves, one is what the other is not, **One is the person we believe we are**.
The other, The "Potential Self", (the Shadow Self), **is the Self we believe we are not**. The Shadow Self offers an alternate behavior, a re-action behavior to the conditions a person finds themselves in, rather than a direct primary behavior response.

Due to past experience "The Self", of each gender, develops a mental template, a mental guide for the individual's behavior that is acceptable to both genders and acceptable to the society they exist in. This mental guide or template is used by an individual of either sex to adapt their behavior to different situations they may encounter. Their primary behavior is that of their physical gender, their secondary behavior is the complementary opposite of their primary gender.

The "Self", we believe we are: is "The Self", of our physical body's gender.
The Mind's "Shadow Self", is the complementary opposite behavior of the gender one believes they are not. If physically a person's gender is male, then their "Shadow Self", is female behaviorally. If one's physical self is female, then their "Shadow Self" is behaviorally male.

Both Selves need to learn to live together harmoniously in our mind, and in our everyday interactions with others. This on-going, life-long balancing process, psychologists term **"individuation"**. A balanced character as befits one's sexuality is the resulting aim.
A well balanced person in every day life, **modifies their behavior for the situation they are in**. being gentle when needed, or when called for being more decisive. * A man does not have to become a female impersonator to do this, but merely choose behavior that is more understanding, less demanding. A woman likewise **under some situations** needs to modify her primary feminine behavior.

The Two Physical Genders

There is a tension between opposite physical genders, "They continually struggle to get along together", and when apart, they "long to be with one another". The attraction of these two complementary opposite sexes to be with one another is in part due to the forces of chemical hormones in the body, and part due to the lure of pleasure when sexually intimate with their opposite gender, and yet there is more to it than physicality. Each person has needs in both mind and body, and struggles to satisfy those needs. Three "needs", of each person are:

a) - The need to be Individual, to be one's own self, to have a say in one's own destiny, **to have a measure of control of one's self**, this is very important on the job and off. To be told what **must be** done every minute robs life of part of the adventure and joy of living.

and

b) - The need to be whole, it is a need to not just be half of one's kind. to intimately join with their complementary opposite gender, however briefly and achieve unity.

Both women and men have these two basic "needs". But there is a third need of both genders, both sexes, and that is, to accommodate an equal existence for being individual, and being a unity, together. This third need **must take place** to avoid the two opposite genders from being in perpetual conflict. This third need is to strive for a "Dynamic Equality",

c) - a "Dynamic Equality", is an equal sharing of existence. This **sharing of existence,** brings about a harmonious co-existence, This dynamic sharing is accomplished by alternating who dominates a situation, taking place between the interactions of men and woman, each needs to let the other dominate as needed, make decisions as needed.

The Needs of "The Self"

Here again are the three main <u>physical</u> and <u>emotional</u> "Needs" of an individual of either gender.

a) a need **"To be whole"**. To be complete, unified.
b) a need **"to exist as an individual"**
c) an equality to the **sharing of existence**, to bring about a harmonious co-existence.

To satisfy these three "needs", a continuous adjustment in attitude and behavior is necessary on the part of both sexes. The effort can produce harmony, Sattva.

All three "needs", or desires can be understood as forms of "Love".
Percy Sledge sang "When a Man Loves a Woman", released in 1966, A song many men can relate to, a "soul", masterpiece. (See also "Love", in the Glossary, page 173.).

* - Tim Conway as gullible, innocent Ensign Parker on the TV show McHale's Navy, was demonstrating one aspect of macho-ism to make sure he could be a manly man and an authority figure to be taken seriously. Mr. Parker standing alone near the PT boat dock, to reassure himself that he could give orders with a manly authoritative voice, gave his gut a pat, and did his

best to let loose "The Macho Animal", "The Beast Within", sounding off with a "Wooo-Ahhhhh" from way down there. Tim Conway I would term not just the people's comic, but "The Comic's comic", he could add an improvised piece in the midst of a planed on-going performance that even "cracked-up" his fellow actors who would try mightily not to burst out of character laughing. Conway's impersonation on a "McHale's Navy", episode, of a Japanese soldier picking-up his mail in camp was a beauty.

Does The Cosmic Ocean, the universe, have a "Complementary Opposite "Twin Self"?

The Cosmic ocean is **"some-thing"**, it is real, it exists, it's complementary opposite self would be, a "**no-thing"**, a thing with no existence. Considering that, we can say:

Universe has no Complementary Opposite "Twin Self ", Outside Itself. All Duality is Within the Universe.

All pairs of complementary opposites are within the universe. Some of the universe is real as on-going activity, that part is an **"Actuality"**. The other part of the universe is real as a **"Potentiality"**. A potentiality is something **with the ability to react to changing conditions**. In the theory in this book, Efforts forces to change existing "held", conditions cause counter-forces to react with efforts , to keep things as they are.

The Body, the substance of "The Self", is the physical vehicle that transports the forces of nature. The Spirit, energy, is the animator of the Self. The Mind is the regulator, this controller of body and spirit, this "regulator", has in this book been termed, "The Heart of Reality".

The Mind of the cosmic ocean

You may think that because the cosmic ocean does not talk on the telephone or say "I think it is time for lunch", that it has no mind. But reconsider, the cosmic ocean assembles the chemical elements and sustains them. The cosmic ocean holds the planets in their orbits, forms stars, and put you together. Surely to do that takes "Mind".

The same three regulators that are at work in your mind are at work in the vast cosmic ocean, and they have billions of years to "arrange things", even with chance as the innovator. Chance can "try again", and again, Logic can fit the results together, pushing and pulling them into position, until "Bob's your uncle" **, it is the cosmic ocean we experience today, and we can both say "It Lives, It Lives".

** - "Bob's your uncle", British-English, with the general meaning of, "and there you are, or there you have it", "that is it". An example using "Bob's your uncle is after directions to travel to meet someone: "go three miles until you reach a crossroads, turn left, go to the top of the hill and "Bob's your uncle", (you are there), say hello".

"The Fates", are
the regulators, the "means", of control.

a) **For Substance** (the body of the cosmic ocean), **"Chance"**, is the regulator.
b) **For Energy** (the Spirit of the cosmic ocean), **"Choice"**, (Selection) is the regulator.
c) **For Forces** of the cosmic ocean, **"The Cycle of Change"**, is the inevitable regulator.

The goal of the regulation in each instance is **to "strive" for Dynamic Balance".** These regulators of the natural world, are termed **"The Heart of Reality"**, in this book.

a) The "need", to be Whole, **attracts complementary opposites toward one another.**
b) The "need", to be Individual, to be "Self". recognizes differences and separates them.
c) The "need", to Share existence, with one's complementary opposite **is a striving for "dynamic balance"**, by alternation of dominance of each extreme of a pair. So both can exist harmoniously.

The three Fares are personifications of the "means" of regulation of the components of the cosmic ocean. Below are their non-personified equivalents.

a) Chance, chance meetings, chance coincidences, chance opportunities, can bring dissolution, and Chaos, but chance can also bring opportunity for the new. The Hindu god **Shiva,** the destroyer, could be viewed as "chance", personified.

b) Choice, choice is selection, discrimination, choices in the aligning, and matching of puzzle pieces, and the mating of complementary opposites together, to bring Cosmos, (order)**.** Choice can be personified as the Hindu creator god **Brahma**, the bringer of order.
c) The Cycle of Change, and the cycle of Alternation. The cycle, can be personified as the Hindu god **Vishnu**, the sustainer god, The "worker", god.

Some chance "happenings", may be small, but may still be meaningful. I went to a male eye surgeon to determine how damaged my eye was in an industrial accident, and if there was a steel fragment in it, while waiting in the small examination room, through a thin door to an adjoining examination room the voice of a female patient reached my delicate ears, it was saying **"But doctor, everything looks so big"**. I began to wonder, how much magnification is enough for new eyeglasses?

Matrix Diagram A of – The Mind

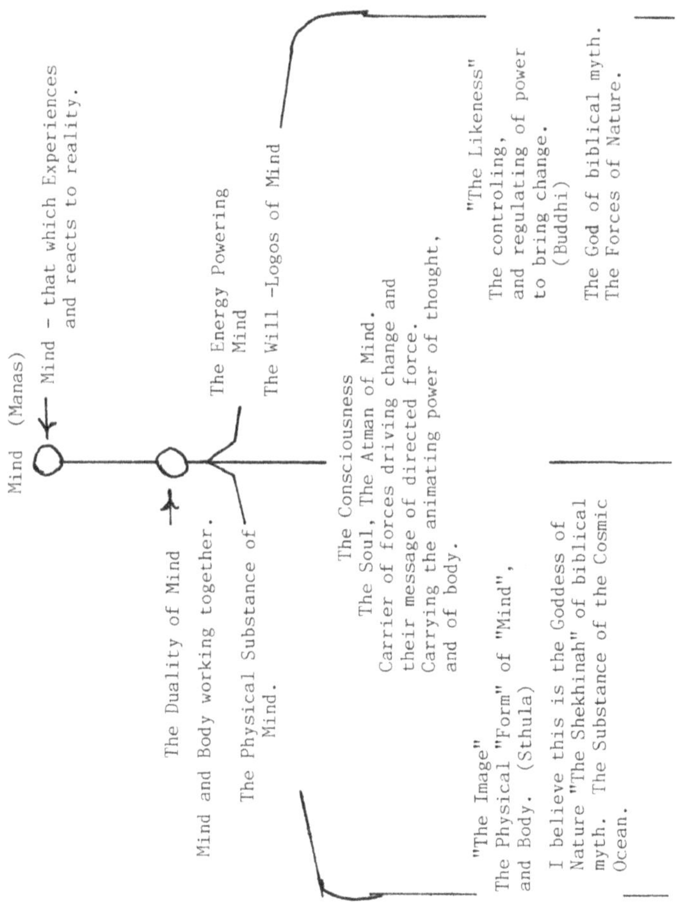

FIG. 14 A The Mind

Matrix Diagram B of - The Mind

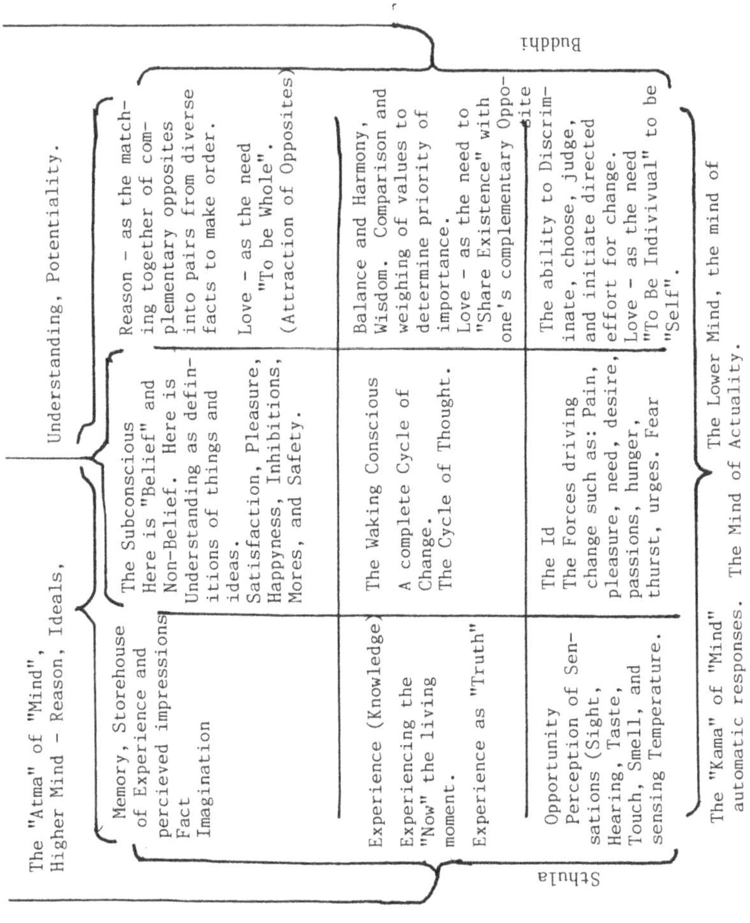

FIG. 14 B . The Mind

When speaking of "The Body of the Mind", in this book, it refers to the **physical material** part of the mind, the brain, nervous system, and sensory parts (eyes, nose, ears, skin, etc. or, the physical movement of electrons and chemicals and the circuitry structure that is "mind", as opposed to the

non-physical ability, the thoughts themselves. **the logic** of the thought process, including any random elements in that process. Together, the physical and non-physical, plus the link between the two, "Consciousness", form the complete "Mind".

In this book the matrix diagram of "Mind", divides "Mind", into three parts, similar to Aristotle's division of the Soul into three parts. The three parts of "Mind", are:
"**Higher Mind**" - The Atma of the mind, the subconscious,
"**Lower Mind**" - The Kama, of the Mind, the Id, and
"**The Mind Between**" - The Waking Consciousness, the Soul, "striving", to achieve it's goal "Dynamic Balance". A "striving" is expressed by the human mind, as emotion).

Lower "Mind" - (bottom row of matrix diagram). The lower mind receives sensations and urges. Lower mind is that part of the mind that controls the largely automatic functions of the body, such as the following:

Growth rate
Breathing rate
Blood pressure
Lymph fluid flow
Heart rate
Appetite

And any similar considerations not listed.

Lower mind tends to separate, to make choices, to decide, to discriminate, to initiate efforts to bring change, gives direction to efforts to bring change.

Higher "Mind" - (top row of matrix diagram). Higher mind, matches diverse data to form match pairs of dualities, compares, weighs values, and sets priorities.
The higher mind's abilities lead to understanding (defining the meaning of terms), and determines relationships of terms. Reasoning is a matching together duality's pairs of extremes. Higher mind tends to bring together and hold as a unity.

The Mind Between - (the middle horizontal row of the matrix diagram is regulated by the inevitable **"Cycle of Change"**.), this central column, is where the forces of higher mind and the lower mind interact, is where force and counter-force interact, and emotion results. As you look down at the matrix diagram of mind, the right hand vertical column is energy regulated by **"Choice",** by logic. The left hand side vertical column of the diagram is substance, regulated by **"Chance",** and in the center vertical column is where the two aspects of force, "need", (effort forces to bring change), and satisfaction (counter-force), are regulated by the cycle of change.

On "Tension", in the Mind

There is a tension between force and "counterforce", tension between the two extremes of a complementary opposite pair. Put in human terms, in the mind, **need** is the on-going effort to bring change. The counterforce is **satisfaction's** effort to remain "as it is". There is a "striving", to achieve "Dynamic Balance", between the two. Striving is the tension between the two, it is expressed as **"Emotion",** by the mind.

In the realm of substance, after all is said and done resolution of substance interacting with substance involves a certain amount of **chance,** in striving for, "Dynamic Equality".
In realm of energy, resolution of energy interactions, usually involves a selection, process, in striving for "Dynamic Equality".
In **the realm of force**, resolution of the interaction of force and counter-force is due to a **cycle of change**, in striving for "Dynamic Equality".

Please note, I said "striving for "Dynamic Equality", not achieving it. The reason being at any one moment one extreme of a duality is always more dominant than the other, this is a condition of inequality, instability. It is only after both extremes of a duality have had their chance to dominate that a measure of dynamic equality exists.

On Dreams

Images in the unconscious mind, Dream images, of sensations, pictures of real experiences stored in the mind, or that the mind assembles from stored information to depict hopes and fears. How real are these dream images to a person? For example: Reliving a terrifying real incident, in a dream. To the dreaming mind, the dream appears as a "reality" being lived all over again, the dream being

so realistic the dreamer may awake sweating and screaming, or reliving an erotic sexual experience. Dreams can also be a product of one's imagination Example: dreaming one is falling backward off a high building, but awakening before hitting the ground in the dream. Another example would be magically flying through the air, but being pursued by an unseen being, and awaking before being caught.
When I dream, most are not in color or at best, a dreamed item will have a faded color. Most dreams if I recall them, appear like a black and white movie, appearing in black, whites and grays. This is apparently true of many dreamer's dreams, the reason for this is not well understood.

Two unique experiences involving "The Mind".

The "special" dream, at the time it may not seem special, not until a week or two later, The dreamer, now awake walks into a building they have not been in before and realizes, the scene of the room they are in, and the scene recalled from their recent dream are the same, I had not been there before, yet this room I am in, is the room from that dream. Such an occurrence may only happen twice in a lifetime, but it makes a fellow think, "How is such a thing possible?"

Before leaving the subject of "The Mind", I will relate an aspect of the "mind" I had intimate personal experience with "Sleep walking". In my younger days while on vacation in the far northern reaches of the state with parents and brothers, I was sleeping in the top bunk of a bunk bed. During the night a loud bang that shook the cabin was heard, this woke everyone except myself. The next morning my father, mother, and brothers told me what had happened in the night. It seems I fell out of the top bunk into the narrow isle between the bed and a dresser with a loud bang, and while still asleep negotiated getting back up into the top bunk and continued sleeping as if the incident had not occurred. In the morning I awoke none the worse for wear, and knew nothing of the big bang in the night. The drop I took was probably about four feet (1.2 meters). I think it is rather amazing the mind has the capability to deal with a situation like that, after all, I had to get up from the floor and climb back into the top bunk with eyes closed and cover-up again. "The Mind", does indeed work in mysterious ways.

Oh! A thing I almost overlooked **"the near death experience"**.
There are persons that have had such an experience, who say, that while in a near death condition they "saw a bright light, and heard a voice, perhaps of an angel or God." Some have had the unique experience of looking down upon their body, an "out of body experience" and seeing others tending to it, including surgeons operating on it and hearing what they are saying. When the person was recovering from surgery they mentioned this to the surgeon, he did not really believe it until the patient repeated the conversation word for word he had had with a nurse while operating. For what I believe is a credible personal account of this, see

author Joseph McMoneagle's book "Mind Trek" copyright 1993, 1997, ISBN1-878901-72-9, published by Hampton Roads Publishing Company, Inc. See especially page 26, and chapter #2. This world can be a bit shall we say "out of the ordinary" at times. My personal experience is limited in the realm of near death experiences. I was operated on in my youth, as a nurse dripped **ether or chloroform**, into cotton wool as needed, the impression I got was on in-hailing the fumes they were heavier than those of gasoline, at one point I regained enough consciousness to hear voices that soon faded. When I was in my early 20's I was anesthetized with what some term **"the twilight zone", sodium pentothal** (the same chemical compound unscrupulous governments use as a "truth serum"), it was used to render me unconscious (this was followed by anesthetic gas(es). I saw no bright light and heard no voices while unconscious. A third time I stood-up on a sunny day, and realized, I did not know if I was standing still or falling over, deciding this was probably not a good thing, I went to see a doctor at an approved, walk in medical clinic, waited my turn for service, then as my blood pressure was being taken, **I passed-out.** I may have looked like death warmed over to the doctor, during all this I felt no pain. When I regained consciousness, I was strapped to a stretcher being given oxygen and nitroglycerine. (At the hospital it was determined the potassium level in my blood, had gone severely low, and my blood pressure, had shot as it were, "out of the ball park", to 280 over120mm.), from my hospital bed, I saw three doctors over near the window discussing something quietly, one glanced over his shoulder briefly at me, with a look I would describe as **"and he is still alive after that happened?"** Still as far as I was concerned, it was close enough to a "near death experience". While unconscious I saw no bright light nor heard any voices. Apparently some persons "do" and some do not, have a more unique near death experience.

The Dreams of George Orr

Ursula Leguin's 1971 novel "**The Lathe of Heaven**", is a science fiction classic, a tale about George Orr, a young man who feared his dreams, because they became fact in the real world, who would ever believe George, when he told them his dreams were coming true. It was adapted into movie form and released in 1980 on PBS Television, This original movie version, with tormented George Orr (Bruce Davidson), A lawyer Heather LeLache (Margaret Avery), who tries to help George, and unscrupulous Doctor William Haber (Kevin Conway), who uses George for his own ends. The Music of this original movie included an excerpt of the Beatles Song "With a little help from my Friends". George Orr's main "friends", trying to help him were The Lawyer Heather LeLache and the state licensed "oneirologist" Dr. Haber. The "oneirologist", was to treat George's neurosis or psychosis that was causing George to use drugs to suppress his dreams.
The story shows what can happen if one person or group says "Our way to live, will be everyone's way". "The Lathe of Heaven", keeps a reader in suspense right up to the last, wondering what will happen to George, Heather, Dr. Haber, and the

situation on Earth. The story is an artistic masterpiece.
In 2002 a new adaptation on DVD (Digital video disc) of the novel was released with the same title. The remake had Lukas Haas as George Orr, Lisa Bonet as Lawyer Heather LeLache, and James Caan as Doctor Haber.

Notes

About Matrix Diagrams

For the most of this book I have spoken of those things which make-up "The Nature of the Existence", Substance and energy, and the link between them. The body and spirit of things we find in the world around us, that drive and human behavior, human needs, desires, and satisfactions. In the full matrix diagrams you will find the relationships of these things and behaviors. Many have tried to organize the reality we experience in the physical world, the nature of Matter, Spirit, the Forms of things, and Life, and so forth.
Thinkers of old have noted there was a "higher" and a "Lower" to their world, a heaven above and earth below, "higher mind" of ideal things, and a "Lower Mind" of earthly needs. The thinkers also noted there were six directions, up, down, and four directions to navigate across the face of the earth. East where the sun was re-born and rose each day, West where it died at nightfall, and the two opposed poles the north and south poles.

Most schemes tried to make sense of and organize information gathered by their designers, but the world is complex in that "though simplicity is golden", problems appeared due to some characteristics shared between whatever basic categories they had found. The stars and "wanderers" (the planets) moved around in astrology, the numbers moved around in numerology, category definitions merge and became fogy. The solution I present here, using the nine square "Matrix Diagram", arrangement, **Fig. 15**, is a way to show component relationships distinctly, as well as a change in the world.

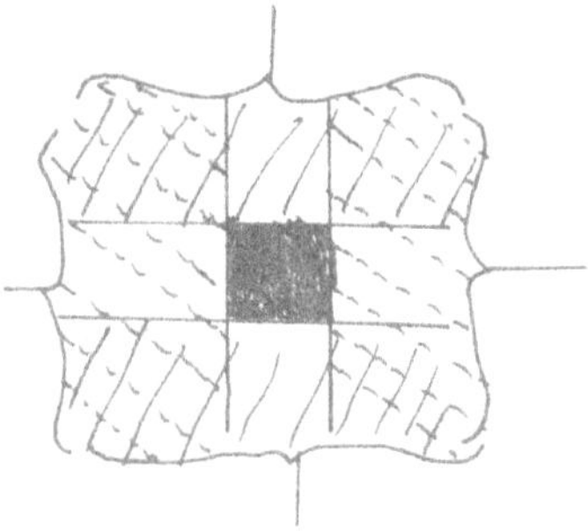

FIG._15_ Nine Square Matrix Diagram

Corner squares are shared by adjacent sides.

In Fig. # 16 page 121, are shown the character of the squares of the top and bottom horizontal rows, these contain the potential aspects of the three basic components of the cosmic ocean, and the actualized aspects.

As well as the primary gender of those aspects, their complementary opposite secondary aspect, is of the opposite gender. It begins with Substance, the mother of all things, the primary gender of substance is feminine.

Energy (the non-substance), is the continual on-going animation of substance. Kinetic Energy has as it's primary gender maleness. Potential Energy, has feminine gender, "Held energy", holder of forms.

Notes

	P rimo rdial Substance	The "Link", Between Substance and Energy.	The Non-Substance Energy
This Row the Diagram is for that with Potentiality and is of feminine gender and reactive aspect.	Space it's primary gender or aspect is feminine. The mother of all things.	Counter-Force it's primary gender is feminine. a reaction force.	Potential Energy it's primary gender or aspect is feminine. Energy held as strain forces in substance.
Tish Row of the Diagram is of either gender or aspect. it is a "Striving", for "Dynamic Equality" "Dynamic Balance". Depending on Conditions.	Chance - Randomness regulated by a C ysle of meetings and misses.	The inevitable progression of a cycle. The Cycle of Change.	A sharing of Existence. Alternation f the aspects or genders of energy. Choice - Selectiveness
This Row of the Diagram is for that which is "Actualized", an on-going Actuality.	Time, it's primargy gender or aspect is masculine. Time lets things "happen". Father Time.	On-going velocity of substance. This "Effort" force to bring change has as it's primargy gender or aspect "maleness". The stress forces that that motion can exert on substance.	Kinetic Energy, the Energy possesed by Moving substance. it's primary gender or aspect is masculine. The animator, father of all things.

FIG. 16 Genders of the aspects of the basic components

Fig. 16 ,is a standard Matrix Diagram, **the Planes or Worlds of Existence, are not shown in this**

diagram for clarity, by leaving out that which is not immediately needed to understand what is shown.
The "planes", or "worlds", are shown in relation to a matrix diagram in **Fig. 17** . The plane of **Atma** would be the three squares of the top row, the "Potentiality", row, **Kama** is represented in the three squares of the bottom row, the "Actuality", row, **Buddhi** is the three squares of the vertical energy column, and **Sthula**, the three squares of the vertical Substance column.
Mind is a rather distributed affair, distributed over the whole diagram. Substance acts like an orb spiders web, sensing all changes occurring in substance, Energy is infused in substance and animates the mind. **Atma** takes care of memory, understanding, and reason functions, **Kama** takes care of sensory functions, and **Sthula** takes care of "fields", and "forms". **Buddhi** takes care of logic functions, selections and choices.

In the vertical center column between substance and non-substance, lies **force**, the link, the messenger, the communicated message, that lets the two extremes, (substance and energy), interact. This communication "Link", is the forces of actuality, and the counter-forces of potentiality.
Note **"Space"** and **"Time" are** in **the Substance** column, in the theory in this book, **both are the Substance of the cosmic ocean.**

The Planes of Existence

Location as concerns a matrix diagram

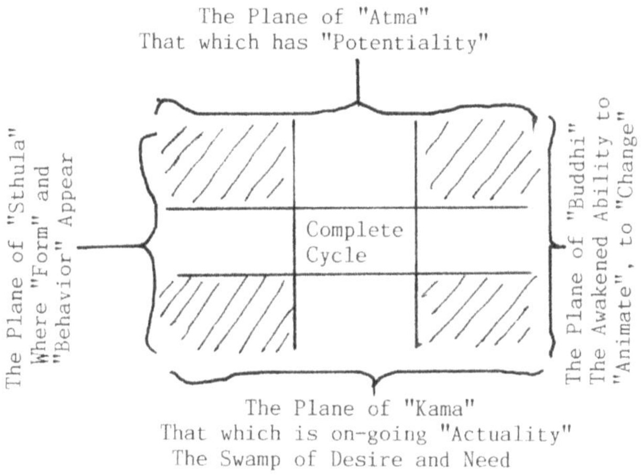

FIG._17_ The squares representing, The Planes of Existence in the matrix diagram.

There are various definitions of these planes of existence, those in the glossary of this book are the ones used here.

These four of the "Planes of Existence" shown, occupy three squares each, one "Plane", to each side of the matrix diagram. The corner squares are shared between two adjacent planes. The square in the center of the diagram is the complete cycle of change. In all matrix diagram have nine squares.

The matrix diagrams in this book, have all been organized along the same lines. The four planes, Atma, Buddhi, Kama, and Sthula, each occupy one side of the matrix diagram (three squares per side).

Two planes of existence, not shown are, **Paran-atma, and Manas**.

The reason being, **Manas (Mind)**, has it's own diagram. "Thought", itself is a cyclic process like digestion, respiration and the cycle of life. Thought involves all three basic components, not just one.
The reason **Paran Atma** is not named on a matrix diagram is, **Paran Atma is "beyond language"**, which is reasonable, it is all the potential that ever was, is, or will be. Paran Atma is eternal potential existence, Paran-atma is the potentiality of the cosmic ocean throughout eternity.

Note: There is always risk of an author being misinterpreted by readers when using terms already spoken about by others, the above interpretations or those in the glossary, are those intended by the author.

The Six "Worlds", or "Planes of Existence", of Metaphysics defined.

The **Atma** (having potentiality), **and Kama** (having on-going actuality), are complementary opposites worlds, as are the pair **Buddhi** (non-substance), **and Sthula** (substance).

1) Atma, - The Plane of Atma is **the plane of Potential satisfactions**, and a world of **Re-Action** to any efforts to bring change.
Atma is potential energy as "attraction", a holding together of an accomplished goal, an achieved wholeness, and Atma is counterforce (inertia). Atma can be seen as feminine in gender, as space substance is the womb of the cosmic ocean from which all things are born.

The complementary opposite of space is time. Time is in the realm of Kama as is kinetic energy, and both are of masculine in gender.
Atma In the world of thought, is "Higher Mind", the mind of ideals, definitions of things, the place of reason", to make relationships, The ordering of diverse facts, the potential to "Understand", and hold experience, as images in "Memory", the "Storehouse of Experience.
Memories" let one imagine "What If this or that was so!", in Atma, is the goal that the lower mind is striving to achieve. In Atma are forces of an **individual's inhibitions and society's mores.**
Caution: here, The World of Atma, is part of the complete self, not the Atman, The complete Self.

2) Kama, - The Plane, of on-going Actuality. Kama, is the realm of **"Actualized", desire and need,** a "Swamp of Desire". In Kama are the continual needs and desires of body, and mind, these can bog down a person. In this sense Kama is a morass,of insecurity, and confusion.
Kama is, a plane of on-going forces with a directedness, either physical or mental. Time substance, is in the realm of Kama, and provides a time "to be". (space substance in Atma, and provides a place for things to exist.). Kama can be considered primarily of masculine gender in behavior, as it's forces actualize efforts to bring about change, they separate, repel, and discriminate. In the realm of Kama, Kinetic energy, is unleashed to deliver motion to substance. In this theory, time is an actualized characteristic, or aspect, of primordial substance. Time is always on-going. In human terns Kama is the mind and body's

needs and desires making themselves known to **"The Lower Mind", known to the Id,. Kama is effort** to satisfy needs. Kama, is a striving for satisfaction by on-going automatic efforts of the Lower Mind, to regulate the body's Heart Rate, Breathing Rate, Growth rate, blood pressure, Peristaltic motions within the digestive system, and the flight or fight response. Kama is where the decisions of the will are initiated. Stated differently if I recall correctly, in director Cecil B. deMille's movie the *Ten commandments*", the pharaoh put the matter of his "will", this way, "So it is written, so shall it be done". **His will as spoken was immediately actualized.** This initiating function, as "judgment" is represented in the matrix diagram's lower right corner square.
In medieval thought Kama was termed **"The Astral World"**, and **to the** Ottoman Sufi, **"The World of Similitude's"**.

3) Buddhi - The plane of, The Spirit, The Will, like the Greek "Logos". Buddhi, is the power, the energy, to animate the physical body, and mind, to make them "to live". Force links energy and substance.
Buddhi regulates the aspects of energy (potential and kinetic energy).
Buddhi is from the word Budh which means "To Awaken", to awaken an ability to regulate change. The aspects of change here are:
a) - Attraction a striving **"to be", whole**, to be one.
b) - Repulsion a striving **"to be", "individual",** to exist as "Self". and

c) - Choice, a **"striving for dynamic equality**, the ultimate goal, mutual existence. Choiceis the "means", of regulation in the realm of energy.

Note: Each basic component of the cosmic ocean, uses it's own unique "means", of regulation, these are covered more fully under the title, **"The Heart of Reality"**.

4) Sthula - Sthula is the physical body. The plane where the **Potentiality and** the **Actuality of Substance interact.** in Sthula, primordial substance animated by energy appears as the shapes of things born, or made, also their characteristics, their behavior, revealed a moment at a time in Sthula. In the world of Sthula, The "Nows", of a born thing become actualized, and In Sthula "The Now", of the present passes into the past, and a now from the future takes it's place to become the "Now", of the present.
Sthula is the living physical body of "The Self", the Atman.
Here in Srhula **Chance** is in control, "Chance", rules.

Note: The spiritual body of "The Self", it is energy, animator of "The substance of the body. Energy is in the world of Buddhi.

5) Manas (Mind), **- The Plane of the Mind, is the capacity of thought.** Mind one might say, has a foot in Atma, a foot in Buddhi, a foot in Kama, and a foot in Sthula. Mind alerts "The Self" to overall conditions, using consciousness to achieve it's goals. It is by mind we receive the impressions of

the world outside of our body, and impressions from within our body. Mind is used to compare, to reason, form thoughts, record the impressions of the senses, including those arrived at by the thinking process. Mind forms ideas, images, and concepts, thus making "wholeness" and order out of diverse received sensations, diverse facts.
"Mind" is organized to satisfy three primary "Needs". The three primary needs of "The Self" are:

a) - **"The need to be a unified whole."** To be as one, together with one's complementary opposite gender (sex).

b) - **"The need to share existence with the ones complementary opposite Self."** To co-operate. Mind chooses which half of one's "Self", needs priority to act, (the gender we believe we are, or the gender we believe we are not).

c) - **"The need to exist as an Individual"** "To be", to act and survive as one's own "Self", physically and emotionally.

6) - Paran-Atma (the world of), - The supreme world of Atma, the highest, or ultimate Atma. **In the theory in this book, Paran-Atma is** The totality of all that is potential. Paran-Atma is Eternal Potential Existence. Paran-Atma is the potentiality of the whole the Cosmic Ocean.

****Caution** – The **Brahman**, referred to in this book, is not a member of the Hindu priestly caste, nor is this Brahman the first god of the Hindu trimurti. The first god of the trimurti, is Brahma, the creator god.

Brahma the bringer of cosmos, the bringer of order.
The other two gods of the Trimurti are:
Vishnu, the sustainer god, the worker god. Vish means "to work".
Shiva, the god of dissolution, the bringer of chaos, and destruction.
The destroyer god.

Simply put, what the "Worlds", of existence do.

Five of these worlds of existence, describe what a living thing is, what it consists of. The sixth world is existence itself, a reality beyond words.

The five planes or worlds that describe what we are, what we consist of are:

1) - We are **our body** (Sthula).
2) - We are our strivings to fulfill **our needs, desires, and deeds** (Kama).
3) - We are **our will**, our energy, and it's regulation (Buddhi).
4) - We are **our experiences, and our understanding** of the world we live in (Atma),

And,

5) - We are "Mind", (Manas), the on-going ability to think, to consider, to imagine, to wonder, and to say "I am, I exist".

Four Worlds of a Living "Self", shown in the matrix diagram.

The worlds, or planes of existence at their simplest.

Atma - The "Self's" thoughts, experiences, imagination. And a vision of the self, and the nature existence.

Buddhi - The Will, and it's regulation. Intelligence, logic, reason, discrimi-nation (being able to tell the difference).

Kama - Need, desire, and a "striving", to satisfy those needs and desires.

Sthula - The substance of a thing, it's physical Body.

The Fifth and Sixth Worlds of Existence.

5) - Mind (Manas) has it's own matrix diagram.

6) - Paran-Atma (the world of), - The supreme world of Atma, the highest, or ultimate Atma. **In the theory in this book, Paran-Atma is** The totality of all that is potential. Paran-Atma is Eternal Potential Existence. Paran-Atma is the potentiality of the whole the Cosmic Ocean.

The Dragons of China and the Pearl of Enlightenment

In Confucian thought, in the beginning there was **The Great Ultimate (T'ai-ch'i)** The originator of all things (The Great Cosmic Ocean with it's complementary opposites). the main opposites in Confucian thought are the Yin (the passive) and the Yang (the active). In the theory in this book we can liken these to the passive (the "Potentiality", the potential reaction to change), that which is "held onto", so as not to be lost. and, it's complementary opposite, the active (the "Actuality", that which is an, on-going effort to bring change). I believe, dragons like humans have a primary active "Self", and a shadow complementary opposite self.

Today Dragons are thought of as mythical creatures, creatures from stories, or are the fearful bamboo, cloth, and paper dragons seen in parades and celebrations. These paper and cloth dragons appear to be part reptile but have a lions head, and part centipede as they have many human legs to propel them along, yet when we face them, entertaining as they may be, deep down, we not only love their presence, but also fear it? Is this feeling due to a "race memory", implanted in the mind of man long ago, or just one's imagination? It is known that once millions of years ago a "Time of Dragons", arose, large, fast, powerful reptiles, roamed land and sea, their fossilized remains are still being uncovered and reconstructed.

The Nine Dragon Scroll

In the year 1244 A.D. during the southern Sung dynasty (960-1229 A.D.) a Chinese civil affairs bureaucrat and artist named **Chen Rong**, completed his painting of nine dragons amidst clouds and mist. The scroll is on bamboo paper, has nine panels which when laid end to end, measure a total length of fifty feet (15.24 m.), there is a direction to reading this scroll, it has a poem at it's end, one of the dragons is breathing out tongues of flame or smoke which point toward this poem at the end. The scroll's action begins on your right as you face it, there you find **"Dragon lies concealed in the deep"**. You may be able to view the scroll on the Boston Museum of Fine Arts internet site, or just ask the internet search engine for "The nine Dragon scroll".

We may never know exactly what Chen Rong thought each scene and each dragon represented, Mister Rong was intoxicated, partly "spaced out", when he did the work. The scroll appears to illustrate steps in a "process", "A Cycle of Change". The stages of development, or trials one faces during life. Again, **the beginning of the scroll is, opposite the end with the poem.**

In a Chinese classic of Confucian literature, The Book of Changes,

(The I Ching), attributed to Confucius and earlier sages, it is written ."**When the dragon flies the time of action is near**". In the centre of Chen Rong's scroll is a swirling vortex, a whirlpool, I see this as **the place of interaction**, it is our world Earth, it is the place where heaven and the underworld interact, or as it has been commented, "It is Heaven and Hell every day", hell being the underworld. Flight across this zone of turbulence can be, perilous. A person's movement through life can be, perilous. I see the dragons as the dynamic powers of nature, the yang or maleness in nature. The mists and clouds, I see as the primordial substance of the cosmic ocean, the yin or femaleness of nature, out of which all is born. Man is part of this natural world. Man's "Self", is subject to the interaction between the yin and yang forces, even with careful planning "chance", can lead to unexpected results.

First I shall describe the dragons of the scroll, as an artist, or poet might, and their location on the scroll as concerns Heaven, Earth, and the Underworld (that which is below Heaven). See Figure # 18 .

The Pearl of Enlightenment
Is The Fruit of Experience

Position of the Nine Dragon Scroll scenes.

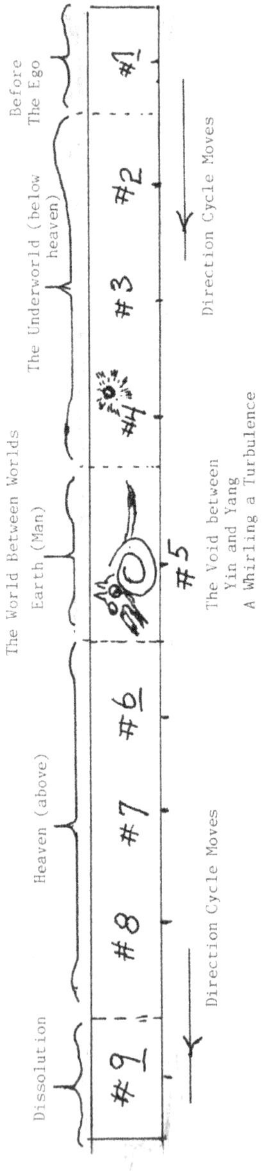

The Numbers indicate the positions of the Dragons in the Scroll of Chen Rong. Read from #1 toward #9.

1) Dragon lies concealed in the deep.
(in the wormb of creation)

2) Dragon appears in the Field
(Dragon acquires Ego "I Am")

3) Dragon Strives to Survive
(goals are needs and desires)

4) Dragon Holds the Treasure.
(pictured as "the night shining pearl")

#5) Dragon Flies Across the Void.
(crosses from the underworld to heaven)

#6) Clouds Follow the Dragon.
(memories of past experiences)

#7) Understanding propels the Dragon
(the dragon on the scroll at #7 has flames from it's mouth pointing the way)

#8) Dragon finds Harmony
(Complementary Opposites Unite)

#9) Dragon Flies To High.
(the old Dragon fades, the cycle ends as new dragon sons and daughters are concieved.)

FIG._18_ Dragon meets a changing World

The struggles of "The Self", to adapt to conditions.
The dragons of the scroll I see as the energy of in the universe, the power to bring directed change, the yang, the active component.
The mists and clouds of the scroll I see as having the quality of Yin, the passive reactive component, molded by the forces and counter forces of our world. The dragon's are continually adapting to conditions.

#1) Fig.__18__, out of the mists and clouds, energy is organizing the form of substance, giving it a directedness, The dragon of "The Self", the ego is forming.
#2) the ego is "born" as it were.
#3) The individual's survival is attended to.
#4) The pearl is grasped by the dragon. The pearl may represent experience, eternal life, a treasure , power, or pleasure. Across the turbulent zone, a zone where the underworld and Heaven interact, a world full of uncertainty and limitations, the world of heaven beckons, a world of certainties, perpetual existence, satisfactions, memories of past experiences, joys, pleasures, and beauties. And so at
#5) the dragon flies crossing over to that perfect world, of ideals, and permanence.
#6) Memories of experiences (clouds), follow the dragon.
#7) the dragon finds Understanding,
#8) the dragon finds Reason, and unites with the complementary opposite of the Dragon and becomes "Whole", complete, or as it might be stated, the dragon and all nature are in harmony. It is the marriage bed of reality, here are conceived the

sons and daughters of the mother and father of all things.
#9) The cycle fades, and a new cycle is taking it's place.

The union spoken of can also be other than a physical sexual union, it can also be a union to harmonize other phases of life. A new harmony between management and the workers, an new effort by an inventor, artist, dancer, musician or laborer to follow their bliss, and achieve a purpose.

I believe the four classes of Dragons represent the personification of the basic components and characteristics of nature. That they represent the same things as the four "Worlds of Existence", of the Aryans, (Atma, Buddhi, Kama, and Sthula). Between these worlds the powers of "held", Potentialities, interact with Actualized on-going efforts to achieve goals.

The four classes of Chinese mythical Dragons,

As they relate to the nine Dragons.

The four classes of Dragon are: (The Celestial Dragon (tien lung), Spiritual Dragon (shen lung), the Earthly Dragon (ti lung), and Underworld Dragon (fut lung). **Heaven here being above earth**. **The Underworld being below heaven**. Earth here is the world between these two worlds. More than one person has commented of their existence on earth, that **It is heaven and hell, every day**. In other words Earth is a place where heaven and hell interact. In the matrix diagram they are directly across from one another, and Earth is on the path from one to the other,

The four realms of the dragons.

Atma, the heavenly world of ideals. **Kama**, that which is below heaven (hell). **Buddhi**, the world of the spirit, a world of power. **Sthula**, the Earthly world of forms and opportunities.

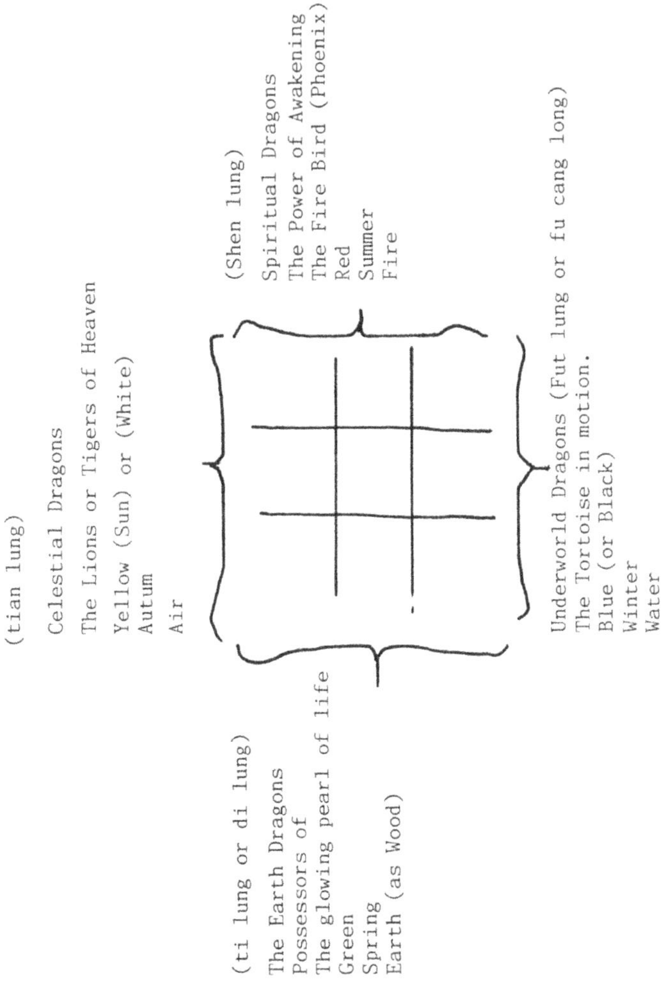

FIG.__19__ The four classes of Chinese mythical Dragons,
As they relate to the nine Dragons.

The Four Types or Classes of Chinese Dragon.

1) **Celestial Dragons** - Helpers of the living, furnishing understanding and the warmth of the sun. They care for the dead, In memories of the

past. Celestial Dragons are "reactors" to change. (Dragons of Atma).

2) **The Underworld Dragons**, The forces of "action" below heaven, the movers of tides, currents and the quaking of the earth. In human affairs, concerns of the forces of need and desire. (Dragons of Kama).

3) **Spiritual Dragons** Regulators and controllers of the winds of change. The awakeners of change, power to attract or repel, the power to balance and harmonize. In human affairs, the power to compare, to judge, and to decide to say "yes" or "no", to "proceed" or "halt". (Dragons of Buddhi).

4) **Earthly Dragons**. Are the Dragons between worlds, between Heaven and the Underworld, Earthly Dragon experiences both worlds, and yet is not totally in either. Earthly Dragon. knows of the heavenly characteristics of permanence, satisfaction, and memory. Earthly dragon also knows "time", ages all "born", things, and that continual adjustment is needed to adapt, not just to it's aging form, but adapt it's behavior to achieve harmony with the natural world. Earthly dragon experiences the joys and sorrows of "being alive", they feel man's condition best of any of the dragon and, are the possessors of the glowing pearl, the pearl of enlightenment, the pearl of a life, of experience and wisdom. (the Dragons of Sthula).

The **Celestial Dragons** who's realm is Atma, The **Underworld Dragons** who's realm is Kama, **The Spiritual Dragons** who's realm is Buddhi, and the **Earthly Dragons** who's realm is Sthula.

We on earth are subjected to forces of each dragon realm, continually adjusting to live in harmony. Some of us succumb to excesses of one sort or another, others of us exist in dire want. Some pray for wisdom whilst others curse the darkness.

Oh! Great One, do we not all wish to know the blue bird of happiness? Perhaps if I help prepare the bluebird a place to live, happiness will come.

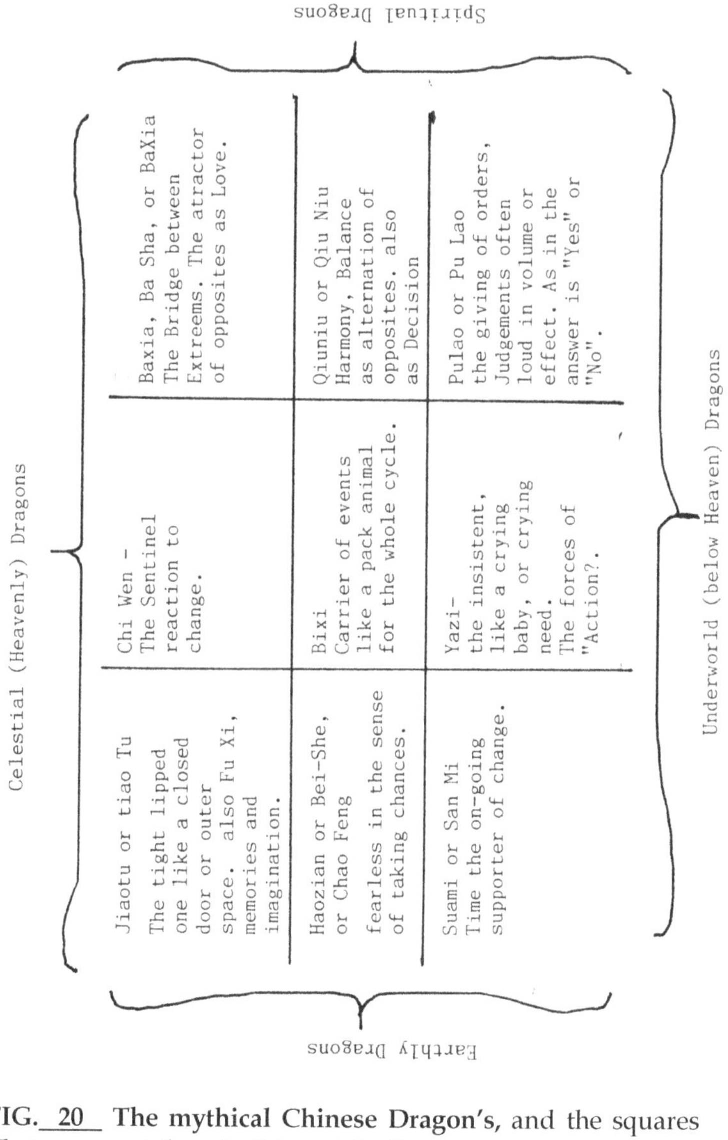

FIG. __20__ **The mythical Chinese Dragon's,** and the squares That represent them in this matrix diagram.

In Figure __20__ The Dragons are personifications of the basic characteristics of the universe, including man's activity and interactions. The **Dragons in the**

corner squares share attributes or qualities of two dragon types. As an example: **Jiaotu** is a **Celestial Dragon** and also an Earthly Dragon.

Please note of the Dragons in the horizontal center row, Jiaotu offers **Chance** as a means of regulating events in the Earthly world of substance, Qiuniu offers **Choice** to regulate events in the spiritual realm, and Bixi offers **"The Inevitable", "The Cycle of Change"** to regulate events.

The Nine Dragons of Chinese Myth

1) Jiaotu or tiao tu, or Fu Xi - traditionally said to be tight lipped, like a closed door or the vacuum of outer **Space**, or like memories and imaginings, as present in the mind or in a closed book..

2) Chi Wen (also known as **Beixi**) - The Sentinel, The Dragon watching over auspicious clarity, Understanding, the way where one sees clearly. In perhaps more human terms as "Belief", perceived reality. Also a protective defensive **Re-action** to change.

3) Baxia, Ba Sha, or **Ba Xia** - The bridge between complementary opposites, in human terms, "Reason" Attraction of complementary opposites, the "Need to be whole" as Love. "Ah! Amore"(Italian), Karlek (Swedish), Liebe (German).

4) Suanmi, or San Mi, or Suanni - Time the on-going supporter of change. In human terms

Opportunity. A thing that can be present one moment and gone the next.

5) Yazi - an on-going actuality like the insistent crying of a baby, the crying out from within of "**Need**", or The **Action** of an aggressive warrior

6) Pulao - The repulsion of likes. The Need to be Individual, The Need to be "Self", The Love of Existing. Survival. A Dragon that can roar, as might someone giving orders or passing on a judgment. Pulao roars "Yes", or roars "No", I wonder if Pulao ever roars "Maybe".

7) Haozian or Bei She, or Chao Feng - In a word "**Chance**". Fearless in taking chances, adventurous, a risk taker.

8) Bixi - like a pack animal. The carrier of events across time and space. The Carrier of the whole cycle of change. **"The Inevitable", "Fate".**

9) Qiuniu - Harmony, Dynamic balance, Wisdom, in essence these are, **a** mutually beneficial sharing of existence, and a cooperation. To an Aryan Hindu, such a harmony is "Sattva", one of the three Gunas (The three

Gunas, or "qualities", are: **Tamas** (the passive, reactive), **Rajas** (the active, the effort to bring change), and **Sattva** (harmony, dynamic balance, sharing).

Notes

On Chinese philosophical thought and change

Before modern times, in China some believed, that from **the "Formless", (the Wu),** would emerge **the "Formful", (the Yu).**
"**Chance**", is aimless (formless) agitation of primordial substance, as a result of which over time, forms can emerge. Also it could be said, that from the non-logical, emerges the logical. In the realm of energy (non-substance), we find that **Potential energy** (held energy), energy of attraction, simply "is", or "is not". But we also find in the realm of energy, **Kinetic energy**, (directed energy), the energy of selection, of separation, (discrimination, the recognition of differences, on the cover illustration of this book discrimination appear symbolically as a wedge separating two parts of a puzzle), this is a form of "**Choice**", as the use of "logic", to arrive at a yes or no, a this or that, or as an acceptance or rejection, a direction to achieve a goal. Choice can give formfulness, as it might to an activity.
The above cycles of change, can as well proceed in the reverse direction. The "Formful", can dissolve back into formlessness. Logic can proceed back into illogic, much as the cosmic ocean can cycle from chaos to cosmos, and eventually, after eons and eons the cosmos can cycle back to a state of chaos.

The Matrix Diagrams and Change.

The three vertical columns (from left to right) are:

1) In the left hand vertical column is represented **That which is changed "Substance",** (or as this

concerns the mind it is behavior that is changed),. (This is the world of Experience, Sthula).

2) In center column are the **forces of change.** In the bottom square are represented **the Action forces,** desire, pain, and fear (in the realm of Kama).
In the top square of this column are represented **the Re-Action forces**, satisfaction, understanding, inhibition. (This is in the realm of Atma).

3) In the right hand vertical column. The power to animate and select the direction change is to take, and initiating the activity of change. (This is in the realm of Buddhi).

The "Shared" Corner Squares

The corner squares are occupied in the matrix diagram by the four basic components of the Cosmic Ocean, The Universe **Space, Time, and the two Energies, held" and kinetic"**, they become all things. **Potential Energy is** both a **"Potentiality" and** a **"Non-Substance"**, an ability. **Kinetic Energy is** both an **"Actuality"**, a thing of on-going change, and a **"Non-Substance"**.
The square between Space and Time, is a result of Time and Space together acted upon by Energy, it is the reality we are exposed to, and part of, dribbled to us an instant at a time.

The theory mentioned in this and two of my previous books

The theory visualizes the universe or an electron, as **four spheres** of **"Substance"** simultaneously occupying the same volume of Space. One sphere of

"Space Substance" is unmovable the second sphere of Space Substance is moveable. The same for the two spheres of **"Time Substance** one is unmovable the other moveable, both occupy the same volume of space.

Note: There is a stipulation however that since these spheres of time or of space constitute "All Space", and "All Time", that they always occupy the same volume of space simultaneously, this means one sphere cannot go wandering off on the others, so **the only motion possible between the moveable and non movable sphere is rotation, a torsion motion**, When movable space rotates however slightly it is termed in this theory **a "Displacement".**

When an object moves through space or time in this theory, the space it is in does not move, only the wave motion, the matter waves", of the objected move as the object travels through the vastness of space.

All change in the universe is due to the application of Energy, Energy as "Force", Force applied to Substance, the Substance of Space and Time. There are three things force can impart to "Substance."

The three characteristics of all wave motion, and cycles of change.

Displacement
Velocity and
Acceleration

In this theory all waves, electromagnetic radiation waves, matter waves, or water waves. **are displacements of the substance, of the Cosmic Ocean.** See Fig. 22 .

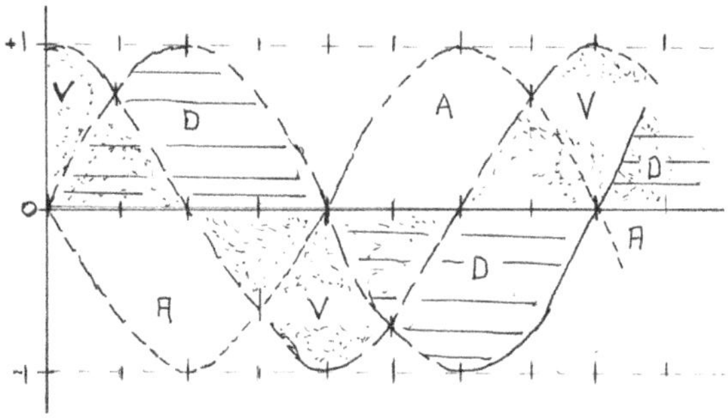

FIG._ 21 _ Wave Graph of **D = Displacement**
A = Acceleration
V = Velocity

Displacement is by rotation, a twist of a moving element compared to the fixed in place stationary element, Fig. 23 . The rotation may be, part of a degree or a few degrees. The moveable space and time substance are as it were, "welded" together and rotate as one. When The Displacement is 'X' degrees the moveable element rotates like a wheel, so the whole wheel experiences the 'x' degrees displacement worth of "Strain" in it's Substance. A diagram of the arrangement FIG.__22_, showing the stationary and moveable elements. The moveable and stationary element occupy the same volume, they are shown separate for clarity of concept. Perhaps you would be more comfortable if we termed the fixed in place, unmoving element "Dark Space" and "Dark Time", This unmoving element is an anchor point for the lines of strain that form when the moving element is displaced. Let us term the moving element "Animated Space, it is the space in which light waves and other waves move. In like manor movable time can be termed "Animated Time" because it is where duration has a time "to be".

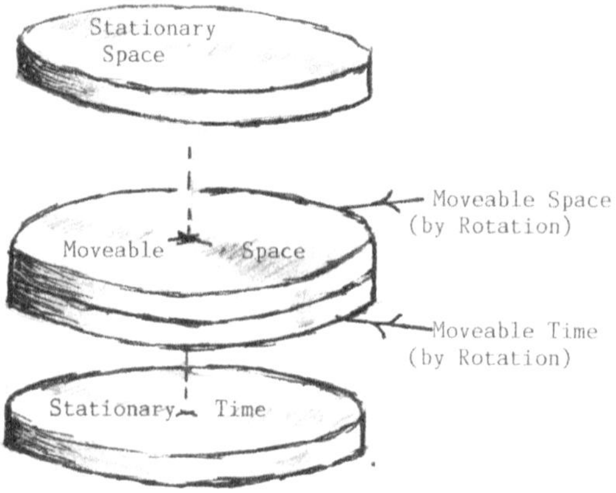

FIG. 22 Displacement by Rotation.

The more Primordial Substance is "Displaced", the slower it's velocity of displacement becomes, like compressing a spring or pushing a car up a hill and then letting it roll back down, at the hill's top the car's velocity is slowest, when the car is let roll back down, it's velocity is highest at the hill's bottom. As displacement of an object becomes smaller, the potential energy it holds, is converting back into kinetic energy (velocity). In a Wave the potential energy is held in a "strain field", due to displacement of substance, in the instance of the car, the energy used to move the car up the hill is converted back into velocity as the car rolls back down the hill.

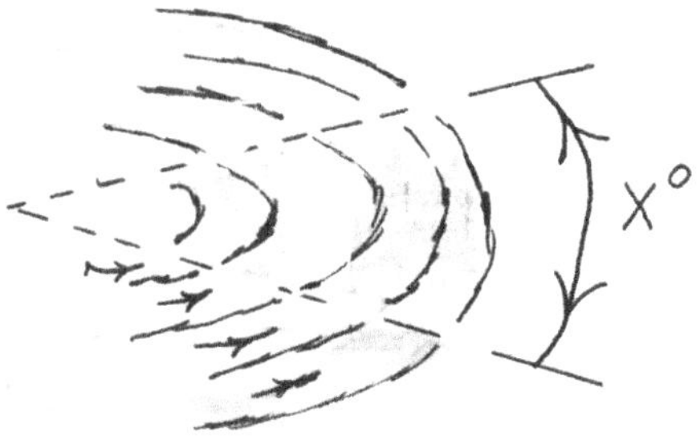

FIG._23_ "Strain" in Primordial Substance is due to 'X' degrees Displacement.

Note: in a torsion wave The direction of Displacement reverses at the peaks of the displacement wave. At the peak of displacement, velocity is zero, and a immediately before and after the greatest displacement acceleration is maximum, (since it is here that the rate of change of velocity, is the most rapid).

As concerns "Velocity" and "Acceleration".- Which came first velocity or acceleration, in other words which is the **"Action"**, and which is the **"Re-Action"?**

My own view of the matter is, the substance of the universe is never still, always moving in some way, so **I see "Velocity" as the "Action"** and **"Acceleration" (a change in the velocity) as the "Re-Action".** That which is accelerated or given velocity is primordial substance it is not rarefiable (made less dense), or compressible (made more dense), there is no place in the universe where this primordial substance of space and time is not, and there is no place outside universe where it is.

Energy is an ability to cause change in primordial substance, and give effort direction. Energy appears in two forms, **Kinetic energy**, that is as moving substance, and as **Potential energy**, which is as a **"Strain"** in substance due to the displacement of substance. This **"strain force"** within substance tends to return substance to it's un-displaced position.

Stress forces drive space and time into displacement. **Strain forces** store or hold the energy of displacement, and give the substance of space and time it's elasticity, it's way to use the stored energy to cause kinetic motion, thus returning substance to it's original un-displaced condition.

So there is no confusion in understanding, in this theory, **Space and Time Substance,** are different aspects of **Primordial Substance, the Substance of the Cosmic Ocean.**

Matter is a special form of primordial substance. Matter is due to energy driving the substance of the cosmic ocean into resonant spherical wave motion. During resonance, energy alternates between potential and kinetic energy.

The Aether or Ether

I consider the **Aether,** or **Ether**, a mythical substance, a substance invented to explain the transfer of forces and energy, The idea that some sort of substance exits however is logical. To avoid the historical "baggage", attached to the terms **aether**, or **ether**, they have not been used in the theory in this book, in there place is **"primordial substance"**, which includes the substance of **space**, and the substance of **time**.

Before moving On

In the theory spoken of above both Space and Time are "Displaceable", both are different aspects of eternal primordial substance. In the theory in this book, **displacement of space,** is a movement from a reference position of (the moveable half of space), by a twisting motion, compared to the unmovable half of space.

Fig.__24__, this is the standard way to graph a sine wave in electronics etc. This graph shows, moments of time elapsed, along the horizontal axis, with the vertical axis of the graph showing simultaneously the changes in amplitude of the three characteristics of a single frequency wave, as it exists.
On looking at the graph you will find the three characteristics of such a wave as it moves through a series of moments in time substance: are "Displacement of space substance", "Velocity", of substance, and "Acceleration", of substance, This graph shows the amplitude of space displacement over a period of time.

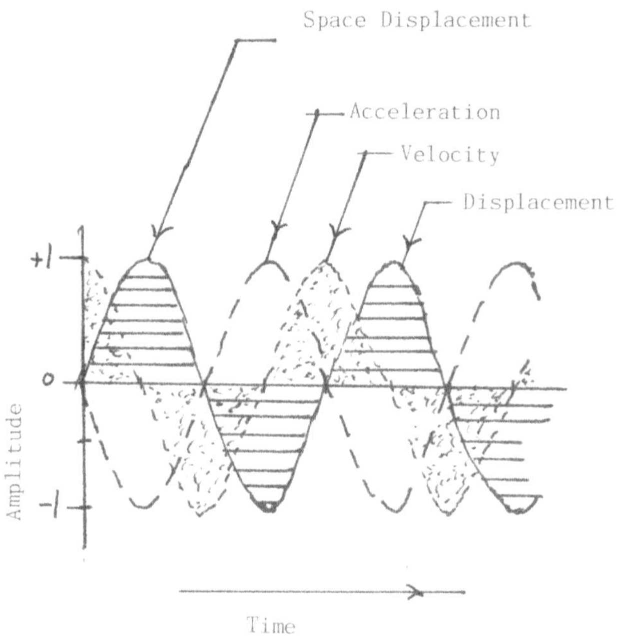

FIG. 24 Time and space substance as usually placed on this type graph.

Caution to students: do not use Fig. 25 , graph information in tests it may give wrong answers to currently accepted views.

In Fig. 25 , **this is a non conventional way to graph a single frequency wave**. This graph shows, distance traveled through space substance along the horizontal axis, with the vertical axis of the graph showing simultaneously the changes in amplitude of the three characteristics of a single frequency wave, as it exists. On looking at the graph you will find the three characteristics of such a wave as it moves through a series of increments of space substance. The three characteristics of such a wave's motion are "Displacement of time substance", "Velocity", and "Acceleration", This graph shows the amplitude of time displacement over the distance traveled.

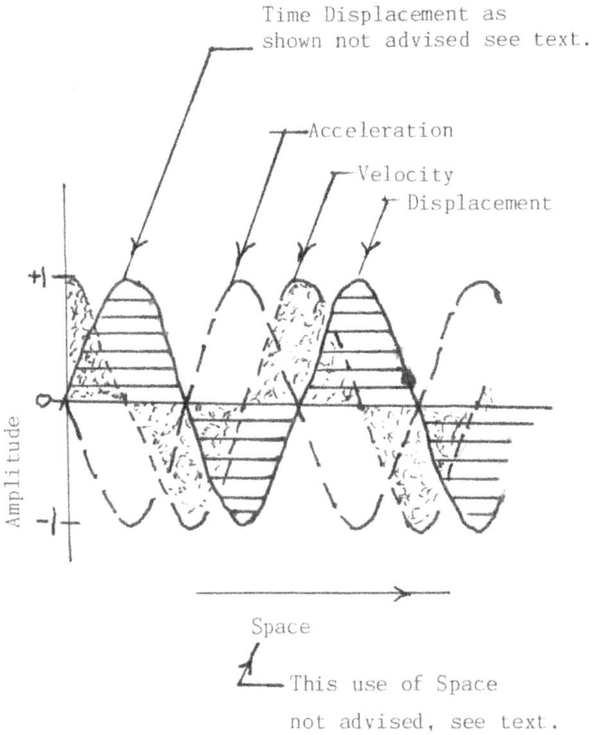

FIG.__25__ On this graph, the positions of Time and Space are exchanged. This is not the usually accepted presentation.

Some science fiction writers and mathematicians might like toying with the above ideas shown in Fig.__25__, in their fiction stories and formulas.

What happens to time substance when it is displaced, does it simply mean time is expended as one moves through space, what other description fits the situation?

Graphic Short hand, Symbols from Long ago

A symbol is a sign, or an emblem. A simple graphic used to represent something, a sort of shorthand way to indicate, a thing, an idea, a thought, or concept. They represent observed characteristics of the natural world, at the time in history when they were drawn.

Some symbols termed **pictograms** are of material objects, such as symbols on modern roadway signs used to quickly indicate, **a curve** in the road, a **hill**, a **crosswalk**, or a **school zone**.
Maps use symbols at times, such as a wavy line to indicate **water**, an envelope symbol indicating a **post office**, or an oval track for **a stadium**.
Other symbols termed **ideograms** have meanings that are not so obvious and generally need to be learned, that is need to be taught to a person, a red cross in a square for a **Hospital**, a letter 'P' in a square for a **Parking Area**. Ideogram symbols can also indicate an activity such as a light bulb above a cartoon persons head, to indicate they have **a bright idea**. A wedding ring band, to represent **unending love,** and not only eternity. Long ago the swastika indicated the **migration of reindeer or birds**. Others saw it as an animated sun releasing energy.

Over the years some symbol meanings have been changed, even reversed. Originally the serpent was neither good nor bad, just another part of nature, in which the **serpent represented rebirth,** the serpent sheds it's skin and as it were, is reborn. Then along came some fellow with his serpent in the garden of Eden story, he could have left the serpent alone and had a talking apple, or a talking rock.

Here are shown some basic symbols that have been in existence a long time, and what I believe their general meanings are.

The Pole, the World Axis, (axis mundi), or "still points" about which one's world turns. A person, a family, or a community, can act as a "still point", or an axis about which events turn.
Examples are: A community celebration that repeats every year, such as dancing around the May pole (in the northern hemisphere this is a celebration of spring, the earth coming back to life to produce her bounty), Religious holidays or festivals, a New Year celebration, or a person's birthday celebration can all act as events about which life moves. A person's job, a source of food supply, health care, are all "still points", about which a persons life can move, some repeat more frequently than others.

The Ring, Rim of a **Wheel,** or **circumference of a Circle, can be seen as** representing time without end, Eternity.

The Spiral - The path to an Origin, or away from the origin.

The Swastika - represents **Restless Energy, or motion,** like that of a swirling radiating sun. The Swastika symbol is quite ancient, it has been found at a Paleolithic archeological site, carved on woolly mammoth ivory, in that instance, the swastika is pictured on the underside of a flying bird's, outstretched wings. It is thought to have been carved as long ago as 12 to 18 thousand B.C. toward the end of the Pleistocene age when the last of it's glaciers were receding. The people that carved the symbol survived by hunting, it was a struggle at the time just to stay alive, the symbol's maker put the effort into making it, because he believed it told a thing of value. **"The whirling swastika symbol"** pictured on a flying bird's wing, tells of an event, **"the flying of the birds that is repeated each year, "bird migration".**

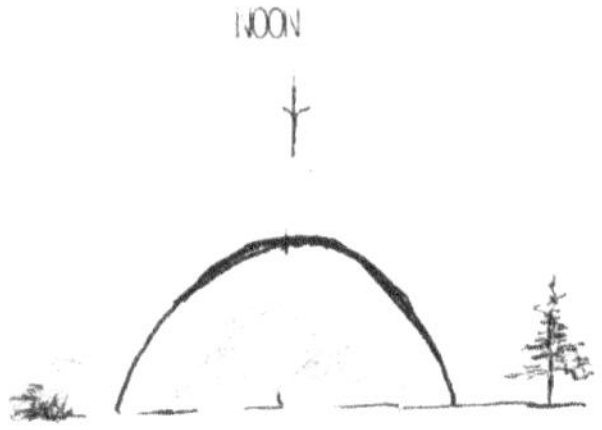

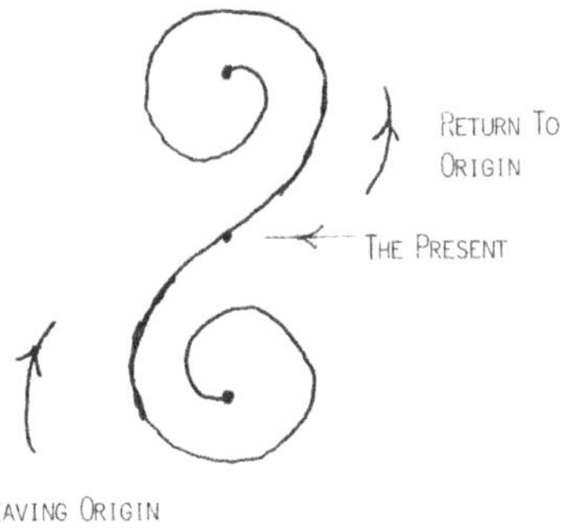

The Arch, To indicate the travel of the sun across the sky. Sunrise to Sunset.

Two Spirals, connected together at their outer ends in such a way that the path of on leads away from it's center, but if one continues on the path to the other spiral it leads back to the center, the origin. To represent leaving an origin and then returning to it.

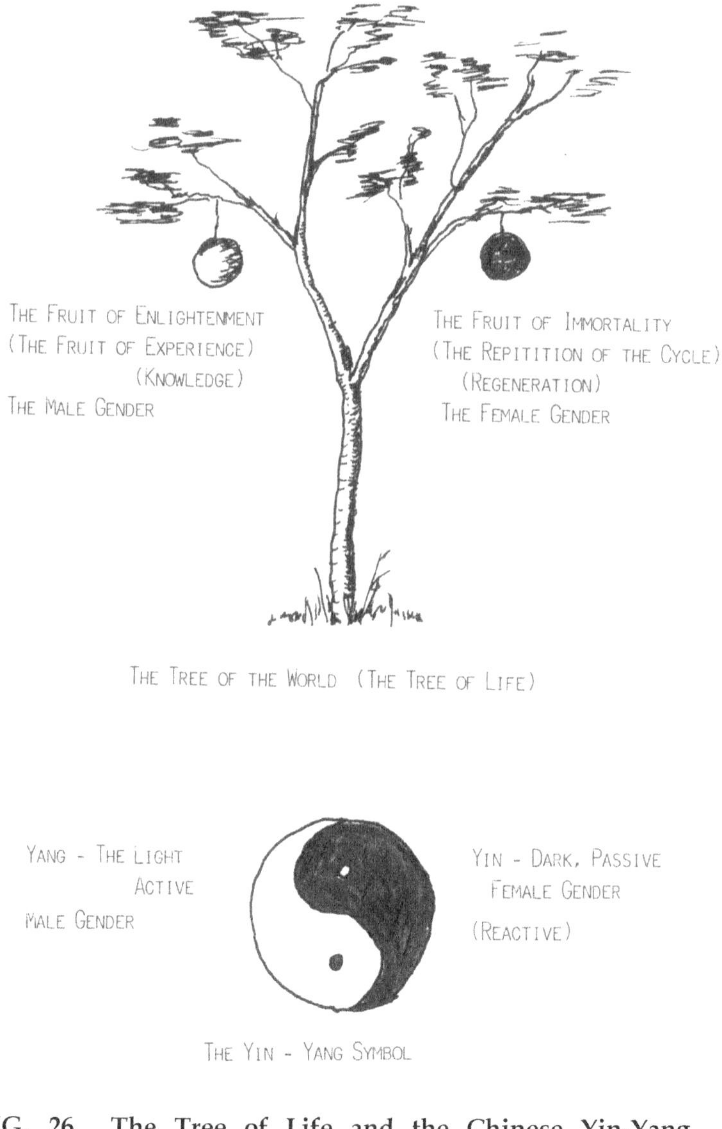

FIG. 26 The Tree of Life and the Chinese Yin-Yang symbol.

Tree of Life - or The Tree of the World, represents, the eternal cosmic rhythm of the cycles of existence in the world of nature, the seasons, and eternal life

as the repeated cycle of reproduction, re-birth, and death.
One of the two fruits of the tree of life, **The fruit of Enlightenment,** defined as being supplied with useful information, informed, or instructed, experience, knowledge, this is **the active male element**.
The other of the two fruits of the tree of life is, **The fruit of Immortality**, is defined as, being exempt from death. This can be seen as eternal re-birth, eternal regeneration**, the reactive female element.** The tree of life has been described as, a tree of instruction in the garden of innocence, it's fruits are those two basic things about which life turns, (knowledge and re-birth), an axis mundi of life.

The Serpent, - can be interpreted as, the power of **eternal regeneration**, much as a snake sheds it's skin, and is as it were, is re-born anew.
The serpent has been given a "bum rap", serpents by and large, are a benefit to mankind.
Eternal regeneration some may view as living as long as the gods, However the goddesses and god(s) are re-born in the mind's of each new generation of men, and each man is therefore the place of rebirth of the goddess(es), and god(s).

The **Yin-Yang symbol -** The existence, of duality within the whole of a thing. The whole of the universe has duality within it, and each individual has that duality within it. As concerns these complementary opposites, there is always a bit of one along with the other. This is shown graphically in the Yin-Yang symbol by the white dot in the black area, and a black dot in the white area. No matter how much one predominates the other is also present.

Ideas do not translate well into images

Expressing an idea with a picture or statue, may lead to a wrong interpretation. The problem is not with the viewer but with the image's ability to transfer the intended idea.

Here are a few examples of images that go against the grain of experienced reality.

1) - A graphic image showing a person with the sexual organs of both the female and the male, to illustrate the idea that the universe has a dual gender. This was done to explain how a single parent could birth children. The "children", of the universe are the things and beings in it.

2) - In East Indian art you may see persons with blue skin. This color may, not look flattering but the color blue is used to indicate they are images of goddesses and gods.

3) - An East Indian art goddess or god image possessing two or three sets of arms. Such images need not be taken as physical reality, if one views them as "stop motion photography", where multiple images are superimposed to show what the goddess or god is capable of at different times, the actual pictured god has only one set of arms.

4) - Then there is a sign made by some persons in this modern world, where the tips of the thumb and next finger over, touch to form an 'O', In the United States this sign generally means "O.K", or "that is good". So a woman from the U.S. went on a visit to Italy and found persons she made this sign to seemed a bit displeased. On enquiring why they were displeased, the explanation she got was, in their part of Italy that sign meant "You're an Asshole." This jolted her into the realization that the same sign in different cultures can mean different things.

About the glossary's definitions of words

Some of the words used in this book may have more than one meaning, more than one definition, **the definitions given here, are those intended by the author** and as such it is hoped they will be useful to you when working to understand the information in this book, other words are easily misused, as I may have done with the words "**affect**", and "**effect**", their proper use is as follows:

"To affect", something, is **to influence** or induce an action, or deed.

Example: The sun affected the fisherman.

"An effect", is **the result** of an action, motive, process, or deed.

Example: The sun's effect on the fisherman, was to give him sun burn.

Over the years some things have been given more than one name. The mythmaker, that is a story teller may prefer one name, the poet another, the physicist another. Each speaks the language of their trade, though they may be speaking of the same thing.

In the theory in this book the following seven terms are used to refer to the third basic component of the cosmic ocean, **"Child of Duality"**, **"The Connectedness"**, **"The Soul"**, **"The Consciousness"**,

"The Link", and **"The Messenger",** are meant to give one the idea of what the third component is, each refers to the same ability in this book. They are the means of communication between Substance and Energy. They carry the pulse of life in the universe, and in all living things.

The focus of the meaning in each of these seven words is different, with **The Child of Duality,** the intent is to show this third basic component of the cosmic ocean would not exist without parents. In **Connectedness,** the focus is on the togetherness, the oneness, the continuum of the whole. In **Soul,** we have the transfer of, the power of life. In **Consciousness,** we have Awareness in the mind of the whole. In **The Link**, the interconnectedness of the parts of the whole is emphasized. In **The Messenger,** we have the carrying of the forces of nature, forces that in various strengths mold our world, and give it behavior.

We sometimes believe we know what a word means, but to define it with other words can prove harder than at first thought.
Whole books have been written about the meaning of some of the terms and things being here defined, for example the three basic characteristics of our universe, Space, Time, and Energy. What really are they? One author wrote a whole book of 191 pages talking about "Time", it is an excellent book, but at the end of it one might say he was at a loss for words to say what "Time" really is.

Be this as it may, words help us to understand, and as such are good, but they can be a poor substitute for experiencing fully the real thing, they are however the best I have to offer you and are described in the glossary definitions.

Glossary

Words here defined are listed alphabetically

Acceleration - Acceleration is the rate at which velocity changes. An increase or decrease in velocity is termed acceleration. A decrease in velocity is sometimes referred to as a "deceleration". Put in other words, acceleration is a change in velocity divided by time elapsed. - **There are however two ways acceleration can occur, and they are:**

Straight Line Acceleration - An object having a velocity increase or decrease while traveling in a straight line. Notation of straight line acceleration is given as: as distance traveled per second and the rate of change in velocity per second. Example **Meters per second, per second** increase (or decrease) in speed, often written as meters/sec. to the second power. This latter does not mean to algebraically square the number of seconds it is simply a short hand way to indicate "per second per second".

Angular Acceleration - An object rotated about a center undergoes continuous angular acceleration. The substance of a rotating car tire experiences angular acceleration, as does the substance of the car, as it turns a corner. A planet experiences Angular Acceleration as it moves in it's orbit. Notation of angular acceleration is given in **degrees of change per second at a given radius**. Or direction change is indicated using a vector diagram.

I will comment on one more aspect of **acceleration**. In the swing of a pendulum, during it's back and forth swinging, there are three characteristics to it's motion, displacement, velocity, and **acceleration**, their simultaneous amplitudes are plotted. over an elapse of time on the graph in **Fig. 27** , below. Velocity can be understood as an **"Action"**, acceleration (a change in velocity), as a **"Re-action"**. And displacement **as the height of pendulum bob's rise and fall** as it swings the back and forth. These three characteristics of the pendulum's motion, are not all "In Phase", with one another, they do not all reach minimum and maximum amplitude simultaneously over time. Fig. 27 . Shows their phase relationship.

The Wave Characteristic Graph

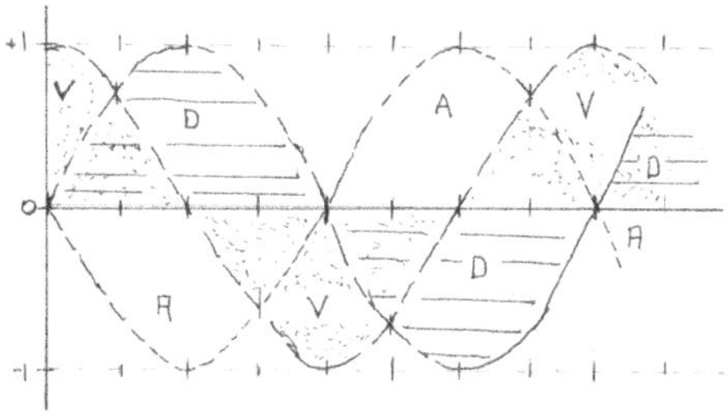

FIG. 27 graph of a pendulum swing (with no losses) Amplitude is the vertical, Time is the horizontal.

D – Displacement of Substance
V - Velocity
A - Acceleration

Action – (Energeia). Action is an effort to bring change, it is on-going motion of substance with direction. Motion delivered from substance to substance is an actuality. In the world of physics, I believe this would be the velocity of substance. An action is also a discrimination, a judgment, the act of separating by decision, **(Energeia)**. These have the Guna (quality), Rajas. The directed effort to move a thing, from one place to another, a motion (effort, force), delivered from substance to substance

Actuality - Actuality is on-going change. Actuality is a thing or activity that is presently occurring, is "on-going", kinetic, has sensible effects, motive (can cause change, movement), is present, current, and existent, a world of pairs of complementary opposites with power held in tension between them, a world of action and reaction. Not hypothetical.

Analogy - Similarity, correspondence, or a parallelism. A thing can be said to be **analogous** when it shows an analogy, **a likeness to something else,** that has similar characteristics or functions in a similar way.

Arbitrary - a choice or selection made, based on random selection or convenience, as opposed to a choice using reason, rules or laws.

An **Aspect** - is, the appearance to the eye or mind, specifically as used in the text, the property, character, or nature of a thing or an ability. example: substance, energy, and force, each have two complementary opposite "aspects", they are as follows in this book:
The two aspects of primordial **substance**, are Space Substance and Time Substance. **The two aspects of**

primordial **energy**, are potential energy and kinetic energy. The two **aspects of Force**, are an **on-going effort to bring change**, a force for change, and the **effort of a thing to resist being changed**, a counterforce.

Atma - (The world or plane of) **Atma**, here in this book, the world of Atma is **the plane of Potentiality**, a world of "held", permanent ideas or forms, and a Re-Action, to efforts to change them. Atma is potential energy as "attraction", a holding together of the accomplished goal of "wholeness", Atma can be seen as feminine in behavior as it is the substance of space, the womb of the cosmic ocean, from it all things are born. Atma is Space Substance which can assume many forms under the influence of energy. (Time Substance is not in the world of Atma. Time is in the world of Kama..

Atma In human terms. **"The HigherMind"**, is a world of ideals, and longed for, and striven for **Satisfactions,** which some would term Heaven. Each person's feelings and satisfactions can be different and so there can be many versions of "heaven". In Atma are forces of an **individual's inhibitions and society's mores**, these are a resistance to change. As concerns the "Mind", Atma is the potential to understand", to "Reason" to relate ideas. The mind can store "images", of experiences in memory, these can let the mind Imagine "What If this or that was so!".

Caution: - The World of **Atma,** is not The **Atman,** "The **Self**".

The Atman - The eternal "Self", the eternal "Soul", of the individual and The eternal Self of all things, The "Soul" and the "Over soul", The Consciousness of Mind, and The Consciousness of the universe, as a whole and it's parts. The Atman has been said to be "The immortal "Self", "The True Self", "The Breath", **The Atman,** the Soul, carries the result of interacting force and counterforce. Carries the message of the aspects of force within substance. Since these forces are always present and active, they are eternal, immortal, they link substance to energy.
To clarify what the self consists of, **The Atman as "The true Self"**, whether The supreme Atman or the Atman of an individual, is the unity of **three basic components of "The Self", these are:**

1) - **The Body of the Self** (Substance),
2) - **The Spirit of the Self**, (the power to animate substance).
3) - **The Mind of the Self,** (the three regulators of the Self, resolving the needs and desires of a person),

There are, so called "lower", and "higher", aspects of each of the basic components, that make up a whole "self".
The Atman, "The Self", has a **"Connectedness",** an innate characteristic of the oneness of substance, and a continuousness of energy within substance. Substance and energy exist without voids or disconnections, this allows the flow of the aspects of force, the flow of the forces of nature, lets the conditions they bring about to communicate, to be "felt", by one another. **There is no action at a distance in this theory.**

The Atman as **"The Consciousness",** is an awareness of the body to conditions within itself and an awareness of conditions affecting the body from outside. The Consciousness , connects thoughts, connects data, and connects ideas.
Caution: Here, The Atman (The Self), is not the "World of Atma".
In this book, The Atman is the whole self. Atma is the "World of Atma", the world of potentiality.

The Balance Point, is the pivot point, the fulcrum where Potentiality becomes Actuality, and contra wise. To achieve a balance is to achieve equality, or symmetry, of opposing forces acting on the balance point.
Dynamic Balance, or "Dynamic Equality", is a point, or place, where the forces acting on it, will be equal and opposite even if they change in magnitude. A flywheel, or a vehicle wheel, that is not dynamically in balance at it's center, will vibrate when spun, the more so the higher it's rotation rate. To achieve dynamic balance of a spinning wheel it's mass needs to be evenly distributed at all points away from it's center. In a harmonic cycle, dynamic balance is achieved, if the three characteristics of the cycle are equal and opposite in direction during each half cycle.
An example of this is an alternating current sign wave, the current flow reverses direction after each half cycle.

Brahman - is described as the non-duality of Vedanta writings, and as **"beyond language"**, which is reasonable as it is all that ever was, is, or will be. It is Eternal Existence, as a singularity.

However it is also said Brahman is **"Kham Brahm"**, which translates as **"All is Braham".**
This allows, that **within the singularity** that is Brahman, **there can exist duality.**

Buddhi - The world or plane of Buddhi, Buddhi is from the word Budh which means "To Awaken", awaken the ability to bring change with direction.
The aspects of change and it's direction are: the need **"to be whole"**, and the need **"to be, individual"**, these are the extremes of existing, between these extremes is a third need, a **"striving for equality of existence of each extreme, t**his need for equality is satisfied by alternation of which extreme predominates over time. It is the alternation of which extreme rules. One extreme attracts complementary opposites, like two puzzle pieces that will fit together, as if they "were made for each other". the other extreme repels complementary non-opposites, such as two puzzle pieces that will not fit together, as if each puzzle piece were saying "I want to be me".
Likewise the extremes, yin and yang alternates which rules, this results in overall harmony, a way to "get along", a striving to co-exist in which there is equal influence over time.

Selectivity is a method of choosing that can be accomplished using logic gates in the thinking animal brain or in an electronic computer, where, a signal is let pass or not pass through a gate.
With the above in mind, Buddhi is the power, the energy, **to reason** (the comparing of facts to reveal relationships), the ability to understand if wholeness or separateness is present. Buddhi

compares and determines priorities of actions to implement change efforts to accomplish a goal. Comparison of distance from extremes, reveals where the mid point is located (the balance point, the knife edge, the pivot point). Then a choice, a decision is made which side is more out of balance, and if any corrective effort need be implemented to reach a desired goal.

Note: In this book I have considered **the undirected animation of substance, the directed animation of substance, and forces and counter forces of nature** separately, as concerns their regulation of change. Each uses it's own unique "means", these are covered more fully under the title **"The Heart of Reality".**

Chaos - disorder, aimlessness, a jumble of semi forms, no sequence or predictable operation. A state in which "chance" is the controller, a state without certainty, The appearance of form and it's duration is unpredictable. Chaos has the **"makings"** of all things, the ingredients but no predictability of becoming them. Chaos is the Complementary opposite of Cosmos, what one is the other is not. Chaos is like, an eraser, that erases the old, to make way for the new. Chaos was once termed "The Abyss", "The Void" because it was where "No-Things" existed, it was also termed "The Waters" because of it's formlessness.

Cold - Absolute cold - A cold level beyond which there is nothing colder, it is primordial substance without motion, no molecular motion, absolute stillness of molecules, quiet. The absolute zero of

the Kelvin temperature scale is when molecular motion stops.
The hottest level of heat would be the inverse of cold. The hottest substance, would be primordial substance driven into frantic motion by energy, substance with, an infinitely rapid frequency of motion. The coldest primordial substance would be perfectly calm primordial substance with no motion, not even wave motion of any sort.

Complementary Opposites - are the extremes in a pair in which one "is" what the other "is not". Each gives the other value. Examples: Female-Male, Substance-Non-Substance, Potentiality-Actuality, Light-Dark. These are the same for all parties concerned. They are termed "objective", in that they are "facts, or characteristics of nature as opposed to an opinion about facts or characteristics of nature.
Right and Wrong are extremes of a different sort, they vary from person to person and from one society and another, as to exactly, what each consists. What one person or society considers acceptable behavior may be taboo by another. These kind of extremes are termed "subjective", they are submissive to the mores of society, or personal views, about behavior.

Confucius (771 - 1123 B.C.) - Grand master K'ung Fu-tzu. K'ung his family name, the tzu part means scholar or philosopher. Confucius taught, Man can make "The Way", "The Tao", great.
To Confucius when speaking of being a gentleman, referred to one's character, not one's birth. The "golden mean" referred to balanced development not just "average" or "moderation". He believed

"Do not do to others what you do not want them to do to you".

Connectedness - A oneness, without voids, uninterrupted substance, uninterrupted energy, uninterrupted connectedness of these two extremes. Connectedness is the integrity, the unbroken volume of substance, or non-substance, The integrity of energy infused in that substance. Before you complain that energy comes in packets, I would comment, energy alternates between it's kinetic and it's potential form, Overall energy is infused continuously in substance. continuous, amplitude may vary, as well as where bunches of aspects exist. "The Connectedness", is Eternal, and is the carrier of the "forces" that animate "primordial substance". Connectedness allows complementary opposites to interact, to communicate, to allow each to effect the other. As used in this book, The Connectedness is, **the togetherness of the cosmic ocean,** that is why any effort to cause separation meets with absolute failure.

If one visualizes the substance of the universe as spanned by a vast spider's web, a web that transmits forces from one part of the universe to another. Change in one part of the web, is felt by the entire web, I would imagine it's strength regulated by the inverse square law. Earthly web building spiders hold a strand of their web, and when it vibrates from an insect struggling anywhere in the web, the spider feels the vibrations and reacts to them, so it is with the cosmic ocean and it's Connectedness. Connectedness, is the same as **The Soul** of a body, or the Consciousness of The Mind. Traditionally the term **Soul** is used in myths and religious works. The term **Consciousness** is used by psychologists, biologists and medical doctors. The terms Resistance, Conductance, and Force, are used in the fields of mechanics or physics.

The Consciousness –is "The Connectedness", "The Messenger", the deliverer of power as forces of nature, and the deliver of information in those forces, throughout the mind.

As concerns the mind, and the body. "Consciousness" has been described as **"self awareness"**. The Self, of mind and body experiencing their existence, experiencing the world within and beyond the "Self". Three divisions of Consciousness have been recognized:

The Subconscious
The Waking Conscious and
The Id

The subconscious is that part of mind that dreams, that believes or not believes. As defined in this book and as placed in it's matrix diagrams, the subconscious includes memory, understanding, (as definitions of terms such as those in a dictionary), and reason, as relationships of terms. Memory and Reason are complementary opposite terms. Reason helps relate and make sense of data in memory.

The Waking Conscious is an interaction of the Id, and the Subconscious. The waking conscious is the mind tending to the business of "The Self", the business of survival.

The Id is the body's automatic response to stimuli. The Id, tells the self this or that needs attention, this or that needs to be satisfied.

To Consider - to reflect on, think about with a degree of care, or caution. To deliberate, ponder, contemplate, weigh.

Continuum - Continuous, An uninterrupted substance or other medium in which there are no voids, no absences, no breaks. **The Substance of the cosmic ocean** is a continuum that may appear in different "Forms", or "Images", due to the action of energy on it. **The energy of the cosmic ocean** is also a continuum, it is infused in the cosmic ocean's substance.

Cosmic Ocean - The Abyss. This Cosmic Ocean is not of the water of our earthly oceans. Earthly oceans consist of the waters of the cosmic ocean that has been formed by energy into "matter". The waters of the cosmic ocean are termed in this book "Primordial Substance". The waters of the cosmic ocean are animated by energy, molded by energy into the earth, the atmosphere, and waters of our world. Energy even forms primordial substance into the fire of the stars.
"All is born out of the Cosmic Ocean, and all fades back into it".
I believe the cosmic ocean is eternal and is the entirety of all that exists.

Cosmos - is order, and harmony. Cosmos a Cosmic Ocean where sequence is present, it is a world of repeated forms, pairs of complementary opposites each giving the other value, When cosmos is present, the extremes of the pair are locked in a cycle where each dominates the scene for a time then the other dominates, a cyclic alternation of domination, this gives a complete cycle of change, a measure of "Dynamic Equality". Cosmos and Chaos are complementary opposites, cosmos is what the chaos is not, they exist as yin and yang do, both are always present, even if one dominates the scene the other is there.

Creation - The use of existing components (substance and energy) to bring greater order, more of Cosmos out of Chaos (disorder). and evolve form. To evolve complex forms and greater diversity of form, from simple forms.

Cycle - A regularly repeated sequence of events, such as the days of the week, the months of the year, the climatic seasons. The vast cosmic ocean is locked in an eternal cycle of change, a cycle of ordering and disordering, driven by an infused and restless power. There is a repeated ebb and flow of the two aspects of the restless power throughout the cosmic ocean's substance. Another cycle is "The Life Cycle", it is the "coming to be" of a thing, it's duration of life, it's reproduction, and then it's fading from existence, leaving behind it's reproduced replacement, which is interpreted by some as a form of "eternal life". **All "born" things, undergo this "coming to be" and eventual "fading away".**

Darkness - Unformed Primordial Substance of the Cosmic Ocean. Darkness is the secret place of God, which surrounds God, like a cloud obscuring the hidden power that is God from visibility. Darkness is, being unable to see, unable to know unable to experience. Light, is being the ability to see, to know to experience that which exists, experience the truth.

Definition - is to "set apart", by identifying marks or characteristics, to make clear the meaning of a word or word group. To define a thing is to distinguish, characterize, or formulate it, as in a dictionary or the glossary of this book.

Destruction - the complementary opposite of Creation. Destruction is what creation is not. Destruction leads the primordial substance of the cosmic ocean to evolve back to simplicity and less diversity.

To Discriminate - to separate, that is realize there is a difference.

Displace - To displace, is to remove from it's usual or proper place. In the theory in this book, the two halves of time and the two halves of space, can be "displaced", one half of each being stationary, the other half of each being moveable. When the moveable half moves from it's unmoved position, compared to the un-moveable half it, by a twisting motion, a partial rotation, it is then said to be "displaced". **In a displaced condition, "strain forces" exist within the primordial substance,** forces that continually try to return the substance back to it's un-displaced state. In the theory in this book this "torsion displacement", of substance may be clockwise or counter-clockwise.

Dissolution (is death, decay) is a dissolving back into that from which a thing was formed, to relax the bonds that hold a thing's form. It is when cosmos (order), reverts to chaos (disorder). A different view of this process is: to separate into component parts.

The Dream Time - To an Australian Aborigine "To be Dreaming", is an "Explaining", of why things are the way they are, usually by the use of story telling, pictures, or actions. The Aborigines had no written language.
A cosmic "Dreamtime", would be a story that tells, of the time of the beginning of things, and a time of the gods

Duality - Dualities are pairs of complementary opposites in the midst of unity. The "Twin Selves" of Egyptian folk lore or the twin selves of psychology are examples of dualities. Some interpret the "Two Lands" of Egypt as a duality of complementary opposites. Substance and Non-substance are a duality, a pair of complementary opposites.

Dynamic Equality, or Dynamic Balance - is a continual striving to achieve the goal of equality or balance. In the theory in this book as long as dualities exist an imbalance exists. At any one instant one extreme of the duality will be dominant and the other subservient. Over time the other extreme may dominate and it's opposite become subservient. This continual alternation of roles is needed to form a cycle of equalization. Still, as said at any one instant one extreme of the duality will dominate etc. This results in a perpetual state of un-balance that continually drives change in ourselves and in the world around us. In the realm of physics, it was once proposed the universe would end in "thermal death", the temperature would even-out all over it and change would stop. Not so in the theory in this book, the two aspects of energy are eternal, they do not wear-out, and substance is regulated by random activity, that activity will not even-out.

Eden or **Paradise** - The garden of God, also termed the **Lower Eden**, a real place that existed on earth once in the area of the Persian Gulf.
Higher Eden or Paradise - the heavenly ideal paradise. The "illa-ah", Wisdom, a form comprising all forms, I interpret this as the "Potential" of satisfaction of needs, and the potential to understand.

The Ego - The "I". The conscious subject of experience, also the ego is, **the assertion of "self determination"**, making choices. decisions, and carrying them out, such as, "I want", "I believe", "I think", "I love", I can do this or that. etc. The "Spiritual Ego" has been called "The Seer of God".

Electromagnetic Force - an **Electromagnetic Interaction Force, a Binding Energy** that holds electrons near the nucleus, attracts electrons to protons, repels electrons from electrons, and repels protons from protons.

The Electron - A basic particle of particle physics, The electron is of negative charge. It has been speculated that a positron is an electron with it's pulsing matter wave, 180 degrees out of phase with that of an electron, giving it a positive charge, and, **that these two particles in various combinations make-up all the chemical elements**. The "mass", of an electron, is 9.109 x 10 to the minus 28th power grams.

In classical physics, zero displacement occurs each time a ball rolls from a hill top, down into a valley, where the ball's displacement is zero and it's velocity maximum, in the swing of a pendulum zero displacement and maximum velocity occur at the bottom of it's swing. **In a standing sound wave,** near zero displacement occurs where there are maximum swings in velocity. This occurs at the open end of a horn (the bell end), this is a velocity anti-node. The lips of the player or the reed, at the mouthpiece end of the horn vibrate, pulsing the air in the barrel of the horn, exciting the air into resonance. The mouthpiece end of a horn, is at minimum change in velocity at resonance, it is a velocity node.

In the spherical standing wave structure of an electron, zero displacement, a velocity node, occurs at it's center, and maximum velocity change is at it's perimeter, This is a velocity anti-node.

Elements - The **ancient elements** of the Svetas Upanishad. VI. 2, are earth, water, fire, air, and aether. **Plutarch** of Athens (Greek philosopher, 4th to 5th century A.D.). mentions these also. They are seen by some to be the product of the unfolding of creation, like steps in the creation process. **The modern elements, the chemical elements,** number some 118. Elements atomic numbers 43 Technetium, 61 Promethium, and all elements with atomic numbers greater than #83 Bismuth are unstable and subject to radioactive decay at various rates. Hydrogen is atomic element #1. Ordinary Hydrogen (also termed light hydrogen), has a nucleus of one proton, and one orbiting electron. Without the orbiting electron the remaining single proton, is called Protium, or in the field of chemistry, the remaining proton is called, the Hydronium ion, with atomic mass of one, it attaches easily to a water molecule giving it a positive charge. When light Hydrogen has neutrons added to it's one proton nucleus it is called heavy Hydrogen. The mass of the chemical elements is greater than that of the particles that make them up, in the theory in this book the extra mass is seen as due to the Strain in primordial substance making up the gluonic fields of each chemical element.

Elohim - gods. This plural name is apparently compounded out of the name of the chief gods of the cultures that surrounded the Hebrews. It is the name used in the biblical genesis creation account, It is a feminine name, feminine gods did the creating. This makes sense, what man ever became pregnant and gave birth to a child.
As a matter of interest: In **Oswald Schward**'s book "The Psychology of Sex" Pelican books, c 1949, it also appears, that man is derived from the female sex.

Emotion - Physical or mental **feelings, or sensibility.** Such as those of: compassion, love, hate, fear, desire goodwill, willingness or aversion. These feelings may bring about changes in the body's hormone levels, respiratory changes, heart rate changes, or changes in the state of readiness of

nerve-muscle response time (the flight or fight response).
Also, **Emotion** - is like a displacement of the substance of thought. Emotion is a striving for a state of resolution, between, the forces and counter-forces of thought, a striving of "need", to achieve "satisfaction of that need". This is a cycle is, a "striving", for dynamic equality.

Energy - Cosmic Energy, Spirit, Prana, is an ability, **an effort that animates substance**, to as it were, bring substance to life. To act to produce an effect. Applied effort is termed "force".
The study of Energy has revealed it has two forms, Energy's femaleness is **Potential energy,** Energy's maleness is **Kinetic energy**. Harmonic wave motion, in this theory, contains, half kinetic, and half potential energy in it's waves.

Caution: Energy and Power were once inter-changeable terms, Today "Power" is defined as energy exerted, that is, "used" over a specified period of time. Power equals Work divided by time taken to do the work. This means, if work needs to be accomplished in a shorter period of time, the same amount of power needs to be used, but needs to be applied within that shorter time.

The active, and passive, aspects of energy.
Kinetic Energy - The work done to overcome inertia. Kinetic energy is dynamic energy, an on-going effort that animates substance. Kinetic energy is, actualized energy. If substance has no motion, it has no kinetic energy. Kinetic energy of a moving substance, can alter" the rate" or "the direction of

motion", of substance it impacts. Kinetic Energy is a measure of the distance a body in motion will continue to move against a constant retarding force. Kinetic Energy can be calculated in classical physics problems as equal to: **1/2mass multiplied by the square of the velocity.** (this formula is not applicable to rotating masses, or "relativity physics" problems, or to so-called massless particles). Kinetic energy in the theory in this book is seen as the push of substance against substance. It is, the stressing of substance, to deform, or to displace it, using velocity to bring change.

Potential Energy - is **"stored energy", or "held energy", "latent energy"**. Potential energy is held in the elastic "strains", in substance that has been stretched, deformed, or displaced. An Elastic substance. holds potential energy, in "strain" forces, or elastic strain field.
In the theory in this book, Potential energy is due to displacement of the moveable Primordial Substance in relation to un-moveable Primordial Substance. Examples of substance with "held", Potential energy, are a stretched rubber band, or in a flexed steel automobile leaf spring. Potential energy can exist in a magnetic field, electric field, gluonic field, or in a gravitational field.

Note: in classical physics problems, potential energy is: **The work done to raise an object in a gravity field,** The amount of potential energy possessed by a mass, by being raised in a gravitational field, equals the amount of kinetic energy gained on letting the object return to it's un-raised position, (minus any losses).

The mores of society contain potential energy, that can act as a counterforce against effort forces that are tending to bring change.

To Enlighten - is to give useful information, to instruct, to supply with insight. To illuminate the mind so it may understand.

Ericapaeus - is: "Action", on-going "Velocity", the power to act on substance.

Essence - The "essence", of a thing, is that which makes it what it is, the substratum that supports it's attributes. Examples are:
The essence of heat, is "motion".
The essence of matter, is "it's substratum, of primordial Substance and Energy. The essence of a person, is their "Self".

Eternal - having no beginning or end. Always in Existence.

Existence - is "To Be", as an actuality in "the now", of time, or have the potential "To Be". Existence is like a ship that conveys us across the sea of time, it is a way to leave the past and enter the future. A thing can have existence as, a material object, an idea, a feeling, an urge, or as a concept.
To Facilitate - To make the bringing about of some action or process easier. Examples: joining two solids "facilitates", the transmission of force from one to the other. or Friendship "facilitates", the exchange of goods. In the theory in this book, the oneness of primordial substance, "facilitates", the flow of the aspects of energy within it.

Fate - examples of the word "Fate" indifferent languages: **wurd** (low German), **urdr** (old Norse), **wyrd** (Anglo Saxon), Fate, is the unavoidable coarse of events. Fate is an impersonal force that predetermines all events. Fate is the cause of destiny. The spoken decree of the gods. The will of the gods. Some would say the cause of destiny is **Need** (necessity), others say it is **Luck** (chance), still others say it is **Love** (a power that can alter many lives). The ancient Greek philosophers had their own ideas about "Fate",

Fate has been personified as **"The Moirai".** The three Fates, described as old wise women, and from the time of **Hesoid** (800 B.C.-Greek poet) onward, were in myth **"The daughters of Night"**.

In Norse and Teutonic Myth they are **Norns**. The dispensers of destiny.

Astrologers predicted a person's destiny, their fate, by the position of stars and planets at the time of their birth, to which Shakespeare (1564-1616) wrote in his play "Julius Caesar" "The fault dear Brutus, is not in our stars, but in ourselves. " Act 1, scene 2, lines 139 to 141. Shakespeare in the play "Macbeth" included three wyrd sisters, **Sisters of Fate**. "Double, double, toil and trouble; Fire burn and cauldron bubble." (Act 4, lines 10 and 11). Probably to be politically correct in that day and age, they were termed "witches" in the play.

The Three Fates Hold the Threads of Destiny,
or put more picturesquely,
The Fates weave the cloth of one's life.

Here **the Fates** are seen as "**The Means**" or method to achieve a result, to achieve "regulation". The regulators of "The Heart of Reality".

Lachesis - Chance, Luck, The random uncertainties of existence. Gamblers use probability to predict results. Probability, is the introduction of "odds" of an occurrence in the field of pure chance, like the odds of rolling a certain number in a dice game.

Atropose - Certainty Inescapable fate against which there was no appeal. For a person the most absolute certainty is their birth, and death. The Inevitable, in the cosmic ocean as in life is, it's changing cycles.

Clotho - Choice. The weaver of the cloth of life, Clotho is the power to choose one's destiny, by making choices.

The Norns - are seen as dispensers of destiny in relation to an individual's experience. Here as positioned in the matrix diagram of **Fig**. 7 , page 60, & page 165, **the main underlying control of that destiny is Chance**, as with the roll of the dice. This does not deny the choices a person makes in life also affect one's destiny, it however claims, that chance, due to random events, such as coincidence, sickness, or accident, underlie the events in one's life. Example: If a person chanced to be at a robbery scene, and were held hostage or wounded, it may hold influence in their future.

The Three Norns, Mistresses of Fate (Destiny), Female Spirits.
Here: seen as **Experience**.

Urd (old Norse), Wyrd (Anglo Saxon), the wisest of the Norns. Here, **"Past Experience"**, What once was.

Verdandi (old Norse), Here seen as: **"Present Experience"**, What is. Experiences of the "Now" can be said to be, "true" things.

Skuld (old Norse), Here seen as: **"Future Experience"**. Experience that may come to be. Change due to random events.

Alternate spellings of the three Norns:**Urth**, **Verthandi,** and **Skuld.**

Notes

	Primordial Substance	The Messenger, Consciousness, Connectedness	Non-Substance, The Spirit, Energy, Ability to bring Change.
Potentiality	The Norn, Urd the past, memory, the wisest of the Norns.	Reaction Forces	Potential Energy
The Fates as a "means" of Control or Regulation.	The Norn, Verdandi Truth, The Present Experience. The Fate - Lachesis Chance, Luck, The present. Randomness personified as a Fate.	The Fate - Atropose, Certainty, The Inevitable personified as a Fate. The Cycle of Change, birth, life, death.	The Fate - Clotho - Choice, Selection, Weighing importance, Wisdom, Balance. The pattern one's Choices Weave.
Actuality	The Norn, Skuld, Opportunity as the future personified.	Action Forces	Kinetic Energy

FIG. 7 , The Fates and Norns.
Fates and Norns as they relate to, the Matrix Diagrams in this theory.

The Field – The field concept can be thought of as **a property of space,** a property of **an elastic substance** or as a property of a particular volume of space.

For example, a magnetic "Field", a holder of Potential Energy, a latent force. An electric field is a holder of kinetic energy. Potential energy "strain forces" in the field can act as a glue, **a gluonic field** to hold (pull together) particles of opposite sign, and repel (push apart) particles of like sign.

A "Field" that can separately and simultaneously transmit multiple "Field", forces, such as light waves crossing through one another, without affecting one another.

In the theory in this book, the condition of primordial substance, is primordial substance that is being displaced by **"stress forces",** or has already been displaced and contains **"strain forces".** Space and time are that primordial substance, it's condition is due to energy's aspects affecting displacement.

Force - A force is **something that impels** a thing, an idea, or action, to bring change. In physics or mechanics a force is **a push or a pull**.

Force is not only a push or pull, it is also a transmitter of information .It is force that molds and moves primordial substance. Force and Counter-force are complementary opposites. **Pressure** is a force, exerted over an area of surface. Example: water pressure is given in kilograms per square meter, or pounds per square inch. **Note:** a spinning flywheel does not have force it has **angular momentum**.

In classical physics:_the force exerted, equals mass times acceleration. This formula can also be written: Force equals mass multiplied by velocity divided by time. **Forces that impel a person to act, or discourage action**, are termed **Motives**. One's needs and desires are motives, when motives are implemented become efforts, strivings, to achieve satisfaction.

In the theory in this book matter in motion can exert **"stress force"**, on non-moving matter and displace it. A stamping machine forms metal by displacing it, causing strain forces in the formed metal. Matter that is displaced contains **"strain forces".**

In the matrix diagram, **the center vertical column, represents the** forces of nature, it's **bottom square** represents on-going efforts to bring change, (in the theory in this book it is of as masculine in aspect). It's **top square** represents efforts resisting change, (it is of as feminine in aspect). **The center square** in this center column, represents the resolution of, force and counter-force, resolution is carried-out by the playing-out of the inevitable cycle of change. The inevitable cycle of change rules in the realm of force. This inevitable cycle, has three characteristics, these occur simultaneously changing in amplitude and direction over the duration of the cycle. See **Cycle** in this glossary for further explanation.

Fortune - Chance, Luck. An unplanned and unexpected meeting or occurrence .Sometimes this is welcome at other times unwelcome.

Gaia - The concept (named after the Greek goddess Gaea, "mother earth"), they believed that the earth itself is a self-regulating living super organism wherein all plants and animals form a system of checks and balances. (proposed in 1979 by James Ephraim Lovelock (1919-______), British organic chemist, and Lynn Margulis (1938-2011), American? molecular biologist).

Gender - The physical sex of a thing, it's Femaleness or Maleness. The most basic pair of complementary opposites in the cosmic ocean is **a) primordial substance which has a female gender,** (the womb of creation). and **b) primordial energy (power), which has a male gender,** (the bringer of form). Each gender has two aspects. The location of the aspects of each gender in the matrix diagram is as follows:

In the top horizontal row, are the **Feminine aspects** of the cosmic ocean, The **feminine aspects** are non-aggressive and re-active behavior, (Tamas behavior), potential to react, behavior that attracts or holds together complementary opposites. The Tamas aspects are: Space Substance, Potential energy, and counterforce.

In the bottom horizontal row of the diagram are the **masculine aspects** of the cosmic ocean, The maleness aspects are actualized on-going active initiative, judgment, directed effort, behavior, to bring satisfaction of needs and desires (Rajas behavior). Will, personified as the God of the biblical old testament. In the theory in this book, the Rajas aspects of the cosmic ocean are: Time substance, (father time), Kinetic energy, and directed force. Force is "The Messenger", between the basic maleness and femaleness of substance and energy. Let us term force, their child, when of female gender it is "counterforce", and "directed force", is of male gender.

The Gluon - An imaginary particle with no mass to carry the nuclear strong force also termed the strong inter-quark force.

God and Goddess - In the theory in this book, **Goddess(es) and God(s) are Nature Personified.** The three components of the cosmic ocean personified are: **1) Substance is Queen**, with two aspects

2) Energy is King, with two aspects. and

3) "Force", is The Messenger, The Child Duality, with two aspects. Force is delivered or carried by "The Connectedness", the oneness, the continuousness of substance and energy. **Believe what you will believe of these things.** However understand, there is only one cosmic ocean, men differ in how they care to see it. Some persons in their mind talk with nature, others Chat in their mind with "The Mother of all things", or "The Father of all things", perhaps asking "Why did this happen?", or "Please help a certain person or persons, to have a better life", etc.

The Grape Vine - A vine that bears grapes as fruit. As commonly used in speech, it can have another meaning, The informal passing of information, news, rumors, or gossip, by word of mouth from one person to another. If a receiver of information asks where the information came from? the reply sometimes heard is **"I heard it through the grape vine". A vine carries** nutrients to the grapes, here the nutrient is information. it is delivered from person to person. In the theory in this book, **"The Connectedness", "The Soul"** and **"The Consciousness"** are **"The Grape Vine" of the Cosmic Ocean,** the information carried (the aspects of energy), is passed along from substance, to substance, in varying amplitudes of displacement. The fruit this vine bears, is sometimes sweet, sometimes not, sometimes gibberish, and sometimes as good as gold, it all however is communication. It is as Marvin Gaye sang **"I heard it through the grape vine"**.

The Graviton - an imaginary particle with no mass, to carry the force of gravity.
Gravity - a force of attraction between objects with mass. In the theory in this book, Gravity, is "strain", due to the displacement of the substance of the cosmic ocean.

Gulf of Ginunga, The yawning gulf, or The Ginunga Gap. - The place of the cosmic ocean's dark formlessness waters, primordial substance in random motion, in Chaos, it's potential to take on "form" dormant, hence like water, a capacity to take on form.

Gunas - see also Prakriti. **The three Qualities of substance**.

1) - Re-action, being passive, termed **Tamas, 2)** - Action, being assertive, termed **Rajas,** and **3)** - Harmony, the alternation of roles in life, **so substance can be all things,** at times passive and at other times assertive, the quality of harmony is termed **Sattva**.

Heart - **Aristotle** (Greek philosopher 382-322 B.C.) said of the human heart, "**The Heart is the seat of the Soul"**. The Heart, as the seat of emotional feeling, as opposed to the brain, as the seat of intellect and reason. Example: one might say "I feel the desire in my heart." The Heart" of any subject is thought of as the place of convergence of something, the center of something, **"The Focus".** Today the human heart is seen as an physical organ of the body, a pump which circulates blood flow in the veins and arteries of the body. In the lungs the blood absorbs oxygen, and releases carbon dioxide (C-O2), delivering the oxygen rich blood, to be absorbed by cells of the body, that in turn release carbon dioxide into the blood which is then returned to the lungs to be re-oxygenated.
Heart's Content - complete satisfaction.
Heart of Hearts - The central or decisive part of something. One's inmost being, actual character, disposition, or feeling.

The Heart of Reality, - of the title of this book is, that which regulates change, or **modifies the flow of change** in the natural world, regulates life. **Heraclitus** (580c-540 B.C.) believed, the "Logos", was shared by all, and is infused in all things, in all substance and energy. Three basic components of the cosmic ocean make-up all things. Each basic component has it's own "means" of regulation. The "means" of regulation, are **Chance,** regulates the encounters of substance with substance., **Choice** (selection), regulates energy, giving effort direction, using discrimination and decision.. and the third "means", of regulation is, the **Cycle of Change**. This Cycle regulates the Forces of nature. These forces deliver energy throughout substance as aspects of energy, and communicate the state of substance back to energy. The aspects of force, move within, and through primordial substance, they pulse primordial substance into and out of **Displacement,** a throbbing flow of change within the substance of the cosmic ocean, it is the beat of "The Heart of Reality". Every cycle in nature, has three characteristics, **1)** displacement, **2)** velocity, and **3)** acceleration).

Heaven - Heaven is both a natural phenomena, and, the satisfaction of what is wished for, a satisfaction of our needs and desires, hopes and dreams.

Heaven as a natural phenomena is an **"actuality",** it exists as the realm of the stars, the moon and the sun that we see when we look up into the sky on a clear night. The floor of that heaven is the ceiling of our world, the Heaven above, the upper world.

Heaven as an ideal place or state, is a "**potentiality"** that exists in the mind of the thinker, this is "**The Heaven Within**", it is the ideal world we would like to experience. Needless to say this potential heaven is not the same in every mind. In fact it could be anything imagined. It is similar to the situation with food, what one person likes another may dislike. The heaven in the theory in this book in addition includes our memories, understanding as, definitions of the words, ideas, concepts, and also as reasoning power, the power to relate diverse data and experiences.

"Higher", and **"Lower".** The use of these terms does not mean one is more worthy than the other.

Each gives the other value, each is what the other is not, each complements the other. The world of the mortal body gives the eternal world of pure thought value and contra wise. Higher would not exist without the lower, they are by that equal in value and equal in importance. Subjects such as "Mind" "Soul", and "World", have been said to have, "Higher" or "Lower", aspects.

The Higher World - the world above the Earth, the world of the sky, the Heavenly world, the world of the moon sun and stars believed by some to be the home of the Goddesses and Gods. Today we know even the heavens of the stars are not static they do change over time, however we can still think of them metaphysically,* as a world of permanence and potentiality, the abode of ideals and the world of perpetual satisfaction.

The Lower World, The World of physical needs and desires. **The world between**, is the world where higher and lower interact.

Idea - An Idea is an impression, a sensation represented in the mind. A thought object, a thing of the mind. Also the complete and final product of reason. **An Ideal** is a thing that is "perfect", "Exact", without flaw, the ideal of a type. example: It is an ideal sphere, in place of the actual thing which has imperfections of state or condition. An ideal can be represented by a word definition or description or drawing. An ideal human is a potentiality, Actual humans have variances of character or form from the ideal. An ideal human would never error, an actual human makes mistakes, some tolerable, some not so tolerable. Some persons are "interesting" due to their non-ideal behavior, others are dangerous. The English like to term persons with certain non-average behavior "Eccentrics". Example: Agatha Cristie's fictional detective Hercule Poirot, could only eat his boiled eggs if the were the same size.

Image - Physical appearance, form, or shape. example: The sculpture, made the statue in her **image**. the form, the shape, given to a substance by energy, some would say, given by god.

Imagination - I believe the Imagination is, the mind's "search engine". The Imagination searches the memory of experience and understanding for already accepted facts to be used as a guide to potential experiences in the future.

Impulse - A push. Impulse equals mass times velocity, or force times time it is exerted,

Individuation - The lifelong effort to seek a psychological balance to one's behavior when confronted with life's situations.
Individuation is a process of using the behavior of "The Self" **we believe we are**, and "The Self" **we believe we are not**. (when appropriate). Life is a two way street (at least behavior wise it should be.) it is not always me, me, me or you, you, you. Sometimes it is a bit of "us", as in using some **understanding** rather than pushing and shoving one's way through life. Dictators demand and threaten. Courtesans smile and have an allure. You or I, Well we just do the best we can. One could term "Individuation" The Cycle of "The Self".

Inertia, other names for this ability to resist acceleration are: **Inertia Force, Inertia Reaction Force, Kinetic Reaction Force,** and the latest "**Mass**". See "Mass" in glossary for further comment.

Instinct - implanted by nature, innate behavior.

Intellect - intellect has been said to be the ability to break-down the whole into it's constituent parts, and so to distinguish their relationships and reconcile their function.

The Jiva - Jiva is from the Sanskrit **jiv** "to live". Jiva is a being's "livingness". The living individual self, the mortal part, the part that is born, lives, and dies. In the universe the Jivas lives are infinite in number. The Jiva, the"life" of the whole, is, the lives of it's parts.

Caution: the next word **"Kama",** is not **"Karma"**.

Kama, - The world or plane of existence of Kama, is a **world of on-going Actuality**, a plane of on-going forces, either physical or mental, with a directedness. Kama can be considered masculine in behavior, as it's forces actualize efforts to bring about change, they separate, repel, and discriminate. Kama is the realm of change. Kinetic energy, is unleashed to deliver motion to substance in Kama, In this theory, Time is an actualized characteristic of primordial substance. **The Plane of Kama,** is **the realm of on-going desire and need,** termed by some "The Swamp of Desire". Kama traps "The Self", as when a person becomes entangled in the demands of continual needs and

desires of life, In this sense Kama is a morass, a state of insecurity, and confusion.
In human terns Kama is the mind and body's needs and desires making themselves known. Also **Kama is effort** to satisfy needs.
Kama was termed **"The Astral World"** (in medieval thought), and **"The World of Similitude's"** (of Ottoman Sufi thought).

Karma - Karma is a deed, a physical or mental action., karma comes from the root "kri" which means support, the idea being everyone supports the universe, put another way "deeds" make the universe what it is. A person's karma is their deeds. One's karma is also the reaction they experience due to those deeds. Good deeds lead to happiness, bad deeds lead to unhappiness. In believing a person is enduring a reaction to their deeds, a person could say of the situation "This is my karma", Also the word Karma then is "Action" (Deed) and "Reaction" as used in this book the word "reaction" also means, that under some circumstances the reaction can be to encourage or aid the initial action as with acceleration. A ball de-accelerate that is slows-down in reaction to rolling up a hill, but accelerates that is speeds-up in reaction to rolling down the hill. The same may be true of a man making an "advance" on a woman, she may **react** by discouraging it or by encouraging it.

Knowledge - To experience is "To know". The biblical "to know a woman", would be to experience a woman during intercourse. Knowledge is also to be conscious of a situation, fact, or concept, that is to know it.

The Lathe of Heaven - A Lathe is a device in which a rotating elastic substance is "formed", or "shaped", by a tool pressed against it. The substance of the **Lathe of Heaven**, is the substance of reality, formed by the animating power of the cosmic ocean. The product of this "Lathe of heaven", is "Life", and "Change". As we sleep, through our minds, flow dreams of forms and behaviors, real and imagined.
In **Ursula K. Le Guin's** 1971 tale titled "The Lath of Heaven", the dreams of George Orr changed reality. George Orr became afraid to dream, began taking drugs to not dream, the state. demanded he get "help" from their doctor to get off drug dependency, George Orr went to that doctor, a Dr. Haber who began influencing George's dreams with his suggestions, and his "Augmenter", device. Dr. Haber tried to "use", George's dream power for his own purposes, ultimately trying to posses the power without using George. The Lathe of Heaven is a fascinating tale, of human needs and desires.

Life - Life in this theory is the activity of "Substance". animated by energy, conveyed by, the consciousness, the soul, "the messenger", the "forces of nature". This definition of life is quite all inclusive in that, all that is animated, that is moving in the universe in a manner of speaking "Lives", in it's way, by this the universe itself can be considered alive, and conscious, as also would be the earth.

Light - Light as an awareness, of more than one's Self. To give light to the eyes, is to give vision to the eyes .In the biblical Genesis account of creation, the separation of light from darkness, on the 1st day does not refer to the light of the sun or the light of the moon. (the celestial bodies do not appear until the 4th day in that account.) The light referred to in the book of Genesis is **The Cosmic Light**, **the ability to bring order**, order to nature, or order to the mind as understanding and reason, "The Logos", of ancient Greek thought.

Likeness - Is to possess the ability to mold, or direct, the livingness, the behavior, of a thing. If a being had the power to create a condition or being, a "likeness" to that being, could do the same.

The Logos - The ancient Greek philosopher **Heraclitus** (580c-540 B.C.) believed, the "Logos", was shared by all, and is infused in all things, in all substance. The Logos is the immanent cosmic governing, regulating, controlling principle in the universe. Logos is present in all things and is eternal. Logos is the "Will", the Spirit, of the universe, the breath of life of all born things. Some would say, Logos is the will of God. An ability, a power to bring order, or release to disorder, the "forms", of born things in the universe.

Love - The desire to satisfy three basic needs: sexual desire, survival desire, and a desire for equality. I believe these three loves, or aspects of love concern one's existence. (See also "need", as love page 109.)

1) the love that draws complementary opposite persons and things towards unity. **The need "To be" Whole**.

2) The love of being, that is the love of existing as an individual. **The need "to be", individual,** and
3) love as, **The need to share existence** with one's complementary opposite. The need to coexist. by alternation of dominance, to bring about **"dynamic equality"**, between the complementary opposites of a pair.

Manas is "Mind", see **"Mind"**

Mass - Mass is the newer name for what physicists once termed "inertia", Inertia is the ability to resist a change in velocity, in a word, acceleration. I might add the eagerness to rename inertia as mass may have it's origins in the fact that physicists still do not know it's cause, the cause of a resistance to a change in the rate of motion. The theory I proposed did give it a cause, the strain in the substance of space, acted as a counter-force to a force causing acceleration. Mass is a measure of an objects resistance to being set in motion_or for that matter resistance to stopping it's motion. This resistance is termed the objects "Inertia". Also Mass is said to be, the " Self Energy" of a particle's bundle of energy.
Mass is an "Energy Cloud". A particle with mass is a resonating "Field of Energy" an energy cloud. A speculation on mass in accordance with the theory I have proposed: - The mass of primordial substance is determined by the velocity, acceleration, and displacement of substance.

Massless - without mass.

Matter - The essence of the material world is it's underlying basic substance, primordial substance. In the theory in this book, **matter, is the Chemical Elements,** which are various combinations of electrons, protons, and neutrons. Matter, is a standing wave phenomena in primordial substance. The electron is an example of a standing "matter wave".

Medium - Something by which a thing is accomplished, carried, or conveyed. Examples are, money carries value, words carry an authors message, music carries a composer's mood, oil carries the pigment in oil paint, a spiritualist carries the message of spirits and delivers it to a person seeking the message. All these are "Mediums" and one more, **Primordial Substance is the means of carrying the aspects of energy within "The Sea of Being" the cosmic ocean,** . Space, and Time **are** the primordial substance of the cosmic ocean. A second meaning of medium, is something that lies in an intermediate position.

Memory - Memory is the mind's storehouse of Experience, it is a holder of facts, and understood definitions of things.

A Messenger - Is the bearer of a message, information, news, or the carrying out of an errand. A messenger could also be one who transfers an ability from one volume of substance to another., such as energy, or force, from a potential form, to an actualized on-going form.

Metaphysics - That area of thought that ponders and considers the ultimate nature of reality, it's origin, it's end, and it's basic components and their relationships-

Metis - Reason, or Intellect. In the theory in this book, Metis. as concerns "reason", is the showing of the relationships of things to make sense of them, to reveal their order so they may be better understood. Some see Metis as **The wise use of knowledge** (experience), represented by the word "Wisdom".

Mind - (Manas is mind**) - The Plane of the Mind, is the capacity of thought.** Mind uses energy to "order", or "disorder" data, facts, or perceptions. Mind, one might say, has a foot in Atma, and a foot in Buddhi, a foot in Kama, and a foot in Sthula. Mind alerts "The Self" of overall conditions, using Consciousness, an "Awareness" of needs and their satisfaction. It is by mind we receive the impressions of the world outside of our body, and impressions from within our body. Mind is used to record sensed reality. Mind compares, reasons, and forms thoughts, and records those arrived at by the thinking process. **"Mind" is organized to identify and satisfy three primary "Needs", of "The Self"**. The three primary inclinations or needs to be satisfied are:

- **a) - The need to be a unified whole.** To be as one, together with one's complementary opposite gender .
- **b) - The need to share existence with ones complementary opposite "Self."** To co-operate. Mind chooses which half of one's "Self", that needs priority to act, (the gender we believe we are, or the gender we believe we are not).

c) **- The need to exist as an Individual** "To be", to act and survive as one's own "Self". A measure of control over one's destiny.

Mind can have the ability to cope with problems and situations, a process termed "Intelligence".
In humans the main physical structure where mind resides is the brain, which acquires information from the five senses: sight, hearing, touch (heat-cold, pain, pressure), smell, and taste. The power to "see" with the mind (to reason) is sometime thought of as a sixth sense, an inner eye, or third eye. **Mind includes** all states of consciousness: **The Subconscious**, **The Waking Conscious**, and **The Id**.

Lower Mind - The mind that regulates the body, automatically, controlling heart rate, breathing rate, growth rate, the blood pressure, digestion, and lymph system circulation, it is the mind of the senses, the mind of needs and desires. The mind of action to bring needed change to accomplish a goal.

The higher mind's abilities lead to understanding (defining the meaning of terms), and determines relationships of terms. Reasoning is a matching together duality's pairs of extremes. Higher mind tends to bring together and hold as a unity.

Higher Mind - Higher mind, matches diverse data to form match pairs of dualities, compares all things to the extremes of their duality, weighs values, holds what are considered goals that satisfy. Mind using memory reasons, understands, and determines relationships of facts, weighs values. The mind that forms ideas and concepts, has a love of wholeness. It is in Higher Mind that, we **believe what we believe**. It is in Mind that we believe what "The Self", is and what it is not.

The Mind Between - is where the higher mind and the lower mind interact, it is where force and counter-force interact, and a measure of dynamic resolution of forces is achieved. The mind between, is where a **"Dynamic Balance"**, is striven for, where satisfaction of need is striven for. The resolving process, is expressed from moment to moment, in the mind, by **"Emotion"**.

Momentum - also termed **"Quality of motion"**. Momentum equals mass times velocity. Momentum is a measure of the time a body in motion will continue in motion against a constant retarding force. Momentum = force x time. t = M/f.

To Mold - to give a distinctive character, form, shape, nature, or consistency. Example: Forces of wind and water molded the landscape. Also to adapt to an existing contour, or behavior. Example: the damp clay was pressed into the mold and took on it's contours.

A Motive - Forces that cause change, that is bring about motion, forces which "Drive", the action are "motive" forces. Example: Hunger is a motive to eat. Greed is a motive to take more than one's share. Gravity is a motive force attracting mass. **To Motivate** is to activate "motive" forces. That is to stimulate an incentive or interest that initiates action. Such as a message of pain from the senses motivates action to reduce the pain. Pain, pleasure, fear, desperation, love, hate, are all "motivating" forces that motivate the mind to direct the body to act, to bring change.

Myth - An invented story of a journey or event of unknown hazy origin usually composed to explain or justify some practice, belief or natural phenomena. Myths can be, experience or dreams, interpreted in mythical (story) form. (a Legend, a Saga.) Some myths help one deal with how to live a humane life. A myth may include a metaphor such as "that is a horse of a different color" which implies that "what is being spoken of, or dealt with, is actually not what it appears to be." Or "We baked in the Sun." implies they were very hot in the sun like bread baking in an oven.

Nature - is the essential character, or constitution of something. All existence is "Nature". The character of space and time is their Nature, The disposition, temperament, or behavior of a living thing is it's nature. One can also say "Second hand reality is the nature of mathematics."

Norns - see Fates.

Order (Cosmos) - created Patterns and Forms,. Repeated arrangements of forms or activity is generally termed order. The repeated "beat" (rhythm) in music is a form of order.

Pain - is the state of experiencing bodily or mental hurt, distress, or annoyance.

Paraclete - Helper, one who speaks on behalf of another. In the theory in this book, "force", is the paraclete between energy and substance.

Paran-Atma (the world of), - The supreme world of Atma, the highest, or ultimate Atma. **In the theory in this book, Paran-Atma is** The totality of all that is potential. Paran-Atma is Eternal Potential Existence. Paran-Atma is the potentiality of the whole the Cosmic Ocean.

Paran-Atman (the supreme "Self") - the supreme Atman is. "The Self", of the whole cosmic ocean, it's body, Mind, and Spirit. Paran-Atman is Existence, as the cosmic ocean. I believe Paran-Atman is the same as Brahman. **Brahman** is described as the non-duality of Vedanta writings, and as **"beyond language"**, which is reasonable as it is all that ever was, is, or will be. It is Eternal Existence, as a singularity. However it is also said Brahman is **"Kham Brahm",** which translates as **"All is Braham".** This allows that **within the singularity** that is Brahman, **there can exist duality.** Paran-atma **"is"**, potential **"existence itself"**. Others have termed Paran-Atma: **The Unity** (of the Ottoman Sufi), **The Ain-Soph** (of Kabbalah), **The Infinite** (of Swedenborg).
Caution: - Paran Atma and, the Paran Atman, are not the same, by definition.

Perception -. The mind receiving direct or intuitive recognition of sensations, ideas etc. Also physical sensation interpreted in light of experience.

The Persona - A person's public personality. The "Front", "Face". or "Mask" a person presents to others. "The Self" one wishes others to see, (appearance or behavior wise, includes telling "white lies", like telling someone they are beautiful when one really believes they are not, so as not to hurt their feelings).

Phanes - light, personified as the first being, Consciousness. Consciousness is here the capability of thought, will, perception, or design. In the theory in this book "Phanes" is the power to bring change, Energy, by another name is The Will, it's two aspects are the power to bring together, to make whole, and the power to separate, to discriminate, to tell there is a difference, to decide.

Phoenix - a mythical bird, **"The Fire Bird",** that when it burns out, is reborn, regenerated from it's ashes, implying it is overall very long lived, perpetually reborn. There is also a long lived date palm, **"The Phoinix Palm"**, this palm is in a manner of speaking is reborn from it's seeds. Egyptians, tropical Asians, and Africans knew of this palm. In this book, one can see all "born things", be it they the Phoenix bird, or Phoinix palm, as the repeating of an inevitable Cycle of change, a cycle reborn over and over, as are plants, and animals, reborn in their young.

The Photon - A photon, is an imaginary particle of zero mass to carry electromagnetic forces, a light quantum, a name for a discrete unit or bundle of radiant energy, an electromagnetic wave packet. Photons are of differing energy content dependent on their wavelength.

Pleasure is a state of experiencing bodily sensual or mental gratification, agreeable sensation or emotion.

The Pollen Path - The path of life. The idea is from the Navaho Indians of North America, which sees Pollen as a source of life, the pollen path then is the path a person takes through life.

Potentiality - A Potentiality is the ability to become different. Potentiality as concerns energy as in the world of physics is the possibility of one "held", form of energy converting into another "held", form. In the theory in this book, when efforts to change "held", views are made, there is the potential a counter-force will react to the effort.

Power - In the field of physics and mechanics, "power", is the amount of work done, and the time it takes to do it. Power equals work divided by time, or to state the same thing in a different manor, Power equals Force applied times distance an object is moved divided by the time taken to do the moving.

Prajna - Consciousness, or Wisdom. The highest soul, That through which "The Self" is perceived, and that outside self is perceived. In this book Prajna is "Consciousness".

Prakriti - Primordial Substance. Primordial Substance is not the modern idea of matter, the "Chemical Elements", **Prakriti is the substratum out of which all things are formed are formed. It is the body of the cosmic ocean**, In this book prakriti is termed primordial substance. Primordial substance is that which is animated by energy, Primordial Substance is displaced by, and brought to life by energy.

Energy gives primordial substance behavior and characteristics. The character of substance has been described as having, three Qualities, termed **Gunas.**

Tamas, passive, a re-action to an effort to bring change to the velocity of displacement of substance.

Rajas, aggressive, assertive, action, the velocity of substance in an effort to bring change.

Sattva, Harmony, a striving for dynamic balance between Tamas and Rajas in the behavior or character of a substance, person or thing, this is accomplished by the alternation of which aspect dominates in it's life.

Prana - The Breath of life. The power that brings an individual, or the whole Cosmic Ocean to life. The energy of the cosmos acting on primordial substance, to animate it. It is the effort forces driving the cycles of nature.

Primordial Substance - That which underlies all "things", their substratum, their essential nature, their essence, a basic component of the cosmic ocean from which all things are formed, (Energy being the forming power). In the theory in this book, Primordial Substance has two

complementary opposite characteristics, or aspects, they are Space and Time, both are **Substance**. **Space is a Potentiality, in that it can assume different forms,** and is seen as being of female aspect. The potential of Space is, all the potential Images or forms it can become when acted on by energy. **Time is an on-going Actuality,** and is seen as possessing a male aspect, giving a discriminating decision or giving a decisive direction to efforts. Substance is never totally "dead", as it is infused with the aspects of energy, it is in a state of restlessness or "held", forms.
Substance reacts to change.

To **Propagate** - is here used in the sense of to spread, to extend, to pass along an effect or action, to transmit, or conduct, a form of wave energy through a substance. A gardener applies the term to increasing the quantity of plants by using seeds or cuttings. **Propagation** to an engineer, or physicist, is the act or action of conducting or spreading an effect through substance, such as light or radio waves moving away from their source.

Propulsion - To push, to propel. having the impulse to cause motion. .

Pulse - A regularly recurrent cycle, such as are repeated surges of blood flow the heart generates, out of the heart into the arteries, and it's return to the heart through the veins, giving a rhythmic beat, a thump-thump, thump-thump, thump-thump. In the instance of the Heart of Reality of this book's title, it's pulse is, the recurrent cycles of change in nature. The throb of a cosmic heart. The most extensive, and all encompassing cycle is, **the rise of order** (cosmos) in the cosmic ocean **to a maximum and the simultaneous decrease in chaos to a minimum. This process is then reversed so that Chaos rises to a maximum as simultaneously order (cosmos) decreases to a minimum.** This comprises one whole cycle, and brings an overall dynamic equality, a dynamic balance to the cyclic process.

Purusha - The eternal Self, The "Atman" of Vedanta. Each individual has a "Self", and the whole cosmic ocean has a "Self", In the theory in this book, Purusha is the body, mind and spirit of a living thing.

The Psyche - The Self of the mind, The totality of the superego, ego, and the id. The conscious and unconscious self. The living self.

Re-action - The response to an action. The counter-force response to an effort force to bring change. A response to a stimulus, response to a medical treatment. A reciprocal or counteracting force (a reaction force).

To **Reason** - is the power of comprehending, understanding, inferring or thinking. To discover, formulate or arrive at conclusions by using relationships of facts. **To bring together facts or suppositions so their relationship can be shown more distinctly.** Example: the matching the extremes of a duality into a pair. The reasoning process allows comparisons and value judgments to be made and also aids in defining word meanings, showing part of the definition of a word is what it's complementary opposite is not. A second definition of **"Reason"** is a cause, or motive for some action or decision. Example: He had a good **reason** for doing what he did.

To Regulate - To alter, to change.

Resistance - To exert a counter force, to an action, **Resistance** may be active or passive opposition. Friction is a common counter force in many mechanical efforts to accomplish a task. Materials with few free electrons (like glass), have a high **Resistance** to electron flow in them, materials with many free electrons (like gold), have low resistance to electrons flowing through them. The opposite of resistance, is the ease of electrical current flow, this is termed Conductivity. Resistance to alternating electrical current flow is termed **Impedance**. Resistance of a material to being magnetized is termed **Reluctance**, ease of magnetizing a material is termed Permeability.

Resonance - is a condition in which energy alternates between it's kinetic form and it's potential form repeatedly. The transfer of energy between it's kinetic and potential form can occur by reflection, an echoing effect. Resonant waves, surge or bounce between reflecting surfaces, and can under suitable conditions form standing waves between those reflection surfaces.

The Self - "The Atman", of Hindu sages. The eternal "Self", the eternal "Soul", of the individual and The eternal Self of all things, The "Soul" and the "Over soul", The Consciousness of the universe as a whole, and of it's parts. **The Consciousness of Body, Mind, and Spirit**.

The **Consciousness, is the interconnectedness of "Mind", and "Body".** The mind's awareness, of conditions within the body and outside the body. **The Self is the entire living person, the whole individual**.

"The True Self", is the unity of **three aspects, or parts, a unified Self,**

1) - **The Body of the Self** is it's (Substance),
2) - **The Spirit of the Self**, (is the power to animate the substance).
3) - **The Mind of the Self,** (regulates the behavior of the Self),

In this book the regulators of the Self are termed "The Heart of Reality".

Unfortunately the abilities of the three parts of "The True Self", are not always put to good uses by humans, King Solomon (king of all Israel 972- 930 B.C.) is said to have commented, "I have seen all things that are done under the sun, and behold all is vanity, and vexation of spirit. **The perverse are hard to be corrected, and the number of fools is infinite."**

from Ecclesiastes ch. 1- vs.14-15.

I believe he was having what is known as a "down day", but you get the idea of what he was trying to say, In my mind I can see him "Crying the blues", going on and on, with a glass of wine in one hand, and with the other hand and arm, gesturing in that Mediterranean way.

Shekhinah - a feminine gender word in Hebrew, meaning, **to inhabit**.
I believe the Shekhinah is the feminine aspect of the cosmic ocean, I believe **"The Shekhinah",** of myth, is The Goddess, queen of the cosmic ocean, "The Shekhinah", is Primordial Substance, The womb of all things. The king of the cosmic ocean, father of all things is Primordial Energy, The God. These two are complementary opposites, Each gives the other value. Worded differently, each is valueless without the existence of the other.
The child of this Queen and King is the messenger, between them, "Force". Each of these three components has two aspects, one with masculine character, the other of feminine character.

Simple Harmonic Motion - Vibratory or wave motion that is of a single frequency, in which displacement is proportional to, and oppositely directed to acceleration, from a position of equilibrium. The position of equilibrium is the position of maximum velocity in a pendulum's swing (the bottom of the swing), this is the same position in a vibratory wave's motion, at which displacement and acceleration are zero, and velocity is maximum. Another example of simple harmonic motion is the motion of a sounding violin string. To my mind a rather synthetic example of simple harmonic motion is that of a point being moved with uniform motion around the circumference of a circle, during half of it's motion it is moving away from it's starting point, and for the other half of it's motion it is returning to it's starting point, In moving completely around the circle the point completes a full "cycle". Such motion, as usually graphed has elapsed time along the horizontal axis and amplitude on the vertical axis. In other words on such a graph, time is a series of equal length moments, against which the variable amplitudes of, displacement, acceleration, and velocity, are compared. See Figure 24, page 142, or fig. 27, page 151.

Sophia, Metis, Wisdom - Comparison of present and past events to use as a guide to make decisions, as to, which coarse of action to take to reach a goal. The result is a "Dynamic Balance", of complementary opposite aspects.

The Soul, - is one of the three basic components of the cosmic ocean, it is an unseen inhabitant, yet the soul's voice is heard by the consciousness as the forces of nature, and felt by the mind as emotion. The soul is, the metaphysical term for the messenger that carries the message of life, the breath of life, the animating forces that make substance to live. In the biblical Genesis story, (Gen. chapter 2:7), the power of God, blew the breath of life into the substance of man and man became a living being. The living "Atman", of Hindu sages, the living "Self". To Egyptians at the time of the pharaohs, the soul was the river of life, the river Urnes along which the solar boat moved.
The Soul is "The Messenger", and the message, that allows the heavenly, and the earthly to interact, allows substance and energy to interact as "Forces", of nature. It could be said **"The Soul", is Connectedness, the continuousness, of substance carrying the animating forces of nature.** These forces in animating substance "mold", the behavior

of substance, they give it character. The Soul also carries the forces of need, and desire, that mold a person's Behavior, mold "The Self". **"The Soul", is the carrier of energy, and carrier of energy's message.** Each living thing has it's own soul, the individual souls are part of the cosmic ocean's soul, "The over-soul", or "All-soul". To the Egyptians at the time of the Pharaohs, the soul was termed "The Ba", the part of a living body that survived death.

Space - is one of the two aspects of Eternal Primordial Substance. In the theory in this book, Space can be "molded", into different forms by energy, with this in mind, **Space is a Substance with Potentiality** that can be shaped into many things and into many fields of force.
Space is place, and when "molded", by energy takes on "form", (a Shape). To describe things in Space, it has usefully been resolved into three dimensions, three directions (each at a ninety degree angle from one another.), these dimensions are used to describe an object's form, relative position and shape. The names of the three dimensions of space are **Height**, **Width**, and **Depth**.

Spirit, (Prana) - Cosmic Energy, an ability to animate Substance, and regulate change. Spirit infused into lifeless Substance, brings substance to life, animates it. **Spirit** is "the power", the power of the gods the Elohim, give to plants, animals and man, so they may live. If the gods one day take back that power from the living, they will die. Death happens eventually to all "created", or "born", things, their individual "swirling's", fade back into the cosmic ocean.. In Physics, The Spirit, is animating energy. The aspects of energy are Potential energy, and Kinetic energy, also an alternation between the two, to bring dynamic balance. In terms of human behavior, Energy's aspects are: **1) The Need "To be" Whole,** to unify with one's complementary opposite. **2) The Need "to be" Individual,** to exist as "Self",. and, **3) The Need to co-exist, to strive for dynamic equality", with one's complementary opposite.** To strive for an existence with the quality of Sattva, harmony. Yin and Yang sharing existence over time in dynamic balance. some state this same idea as, **Love thy neighbor as thyself**, this includes the two selves of the psychologist, "The Self" we believe we are, and, "The Self" we believe we are not, these two selves are as it were neighbors. Love, as energy, **completes** the three aspects of choice, or as it also can be stated, **Love fulfills the law. Energy can be said "to be", Love, is existence**., Paul of Tarsus (formerly named Saul, born between 5 and 15 A.D., died about 67 A.D.) in his letter to the Galatians, wrote in Chapter five ,**"The fruit of the Spirit is love"**.

The "Still Point", of existence - That about which change takes place. The "Still Point", is the focus point of change, much as the axis of the world is that about which the earth moves, or the focus point of one's life, it is that point about which one's life revolves.

The ultimate "Still Point", for a person is their "being", their existence. Life revolves around existence. Life revolves around birth and death, politics, self, family, neighbors, work, religion, war, or "a need to know and understand". A farmer's life revolves around the seasons, a hunter's around game migration. It is **imagined** that if a person

stood at such a "Still Point", they would be, pure existence standing there as the cosmos swirled about them. But existence in the cosmic ocean, is not as quiet and peaceful as it may seem, our existence affects, the existence which we are not (that outside our "self"), and which we are not, affects us. Repeating cycles of change between the two, go on continually maintaining a balance, an overall equality. If a living thing regenerates before extinction, the cycle of existence repeats, that is life.
In astronomy there are what can be termed "Still Points", zones where there is a balance between the gravitational forces of two bodies or more. Some are ephemeral, fleeting, less stable, others, more stable. If a person stood at such a gravitational balance point, they would "float", suspended in space, as if unaffected by gravity forces. These places in astronomy are termed Lagrange Points (named after French mathema-tician **Joseph Louis Lagrange**, 1736-1813). Between a sun, a planet and a moon, five such points in space are recognized, among which, **one is the most stable**. The Lagrange points are only truly still, if the moons, planets, and suns bringing them about are still, which they are not, so exactly where the points are, depends on the positions of the moving bodies producing them. A space traveler resting in a very stable Lagrange point, may as it were, be "sucked along with it, or contrive to stay with it". At a less stable Lagrange point stability deteriorates and one is pulled toward the strongest gravity field.

Sthula - is the world or plane where "Formed Things", appear. The plane where **Potentiality and Actuality of Substance interact.** Sthula, is the plane where the "Potentiality", of space substance meets the "Actuality" of time substance. Sthula is where time and space, meet and resolve into a flow of form, a flow of duration. Sthula is **"The Nows",** of a person or thing's existence. In the world of Sthula, primordial substance, the physical body of a thing is animated by infused energy. In Stula, Substance attains behavior, attains character. In Sthula the state of things "born", appears a moment at a time, is born grows, ages and eventually looses it's ability to retain it's form, it dies. Here in Sthula "The Now", of the present, passes into the past, and a "Now", from the future takes it's place to become the present. Sthula is where space is molded into force fields and shapes. Here in Srhula **Chance** is in control, Chance rules the realm of substance, rules chance encounters, chance happenings, and chance outcomes.

Strain - The condition in an elastic substance subjected to stress forces. In the theory in this book these "strains", in substance, are due to it's displacement. Substance is driven into displacement by stress forces imposed on it. "Strains", in the elastic substance of space appear to us as "forms", Chemical elements", and "Fields of Force", (strain force fields.)

Stress - The non-uniform forces that cause deformation or displacement in an elastic substance. Stress forces bring on the condition of strain in an elastic substance. I believe the substance of space itself is an elastic substance.

Substance - see Primordial Substance, (Prakriti).

Super ego - the subconscious referenced by thought processes. The super ego can act like a warning at the back of the mind, a brake to control the impulses from the Id. The super ego accesses the subconscious which includes a person's inhibitions, and the mores of society that serve as aides to character formation.

Nuclear strong field of force, - "The Nuclear Glue" "the Gluon field" or gluonic field, holds together that which makes-up protons and neutrons. It holds an atom's nuclear fields in order. This nuclear strong field of force is 10 to the 41 power stronger than the gravitational force. A Gluonic Field, can be seen as a localized wave interaction zone, where the waves that make-up particles like protons and neutrons interact and resolve as "The Gluonic Field. See the work of Gabriel LaFreniere,of Bois-des-Filion in Quebec Canada, his work is titled "Matter is made of Waves". This work has been published on the internet.

Tendency - a readiness for a particular action, or in a particular direction. An inclination.

Tension - The dynamic balance between opposing forces. An elastic material, like the substance of the cosmic ocean, is under "tension", when forces are stretching it, displacing it. Also, in the world of art, to a viewer, colors or shapes, can appear to be in tension, opposed in some way. In the world of physics, the tension between high and low electric pressures, (voltages), as when an electric transmission line, is said to be a "high tension line", high electrical voltage within the line, and low voltage outside of it. **Tension** is present within Primordial Substance when it has been displaced, tension within the displaced substance appears as "strain forces". In the realm of the mind, tension appears as emotions.

The Third Eye - The eye of the mind. The so called third eye perceives by reason, by logic, by comparison, by discrimination (noting differences.), using the mind. The other thing the mind sees is "The Self", and that which is "Other than Self". The mind sees ideas, concepts, visions, and things imagined.

A **Thought** - Thought is the arranging of facts and ideas in the mind, also a recalling of impressions, concepts or representations from memory. Consciousness routes the forces of the senses, and need, through a controller in the mind that directs them to be stored, recalled, compared, related to one another, assessed for use, and if acceptable the information is acted upon to satisfy needs, desires, and urges. It is in one's thoughts a one sided conversation with a Goddess or God can be had. This can be therapeutic, and helpful, ask what you will ask and let your mind search for the answer.

Time - Time is one of the two aspects of eternal primordial substance. All consuming "Time", is. said to be **"The great Equalizer"**, in that the weak and strong, the poor and rich, peasants and kings, become equal in the end, all die. **Time** is an **Actuality**, not a potentiality like space. Time is always present, always there, always on-going.

"created things", have a limited life-span, they are born, exist for a time then die. Time is the facilitator of change, without time change could not occur. Time is the elapse, the duration for change to take place in, time, "to be". Time makes way for what is to come after "the now". **Change is not instantaneous**, a period of time is needed for the forces of the universe to increase or decrease. In an analogy, time is like a road down which the vehicle of existence moves, without the road of time the vehicle of existence would remain parked, motionless. A **"Duration"** is a span of time more than an instant, a duration is a series of instants, a series of moments, a duration is an extended moment, a duration of time in which change can occur. I might add with a smile, this example of a duration, "The activity will begin **"Shortly"**, or in a **"Jiffy"**, that is in a relatively short **duration** of time. Events that occur during the same moments are said to be **"Simultaneous"**. Example: six persons starting a foot race at the same moment, as a spectator spilled popcorn, These occurrences all shared the same moment, they occurred simultaneously.

In the theory in this book, **Displacement of time** is like a stretching, a of a rubber band and when the stretching forces are removed, letting the rubber band return back, to it's non-stretched, non displaced condition.

Time that is not displaced, is "Eternity", is "stillness", with no change. **Eternity** is time without beginning or end, sometimes termed "Boundless Time". Fortunately for living things like ourselves, our time is displaced. Multiple displacements exist, some overlapping, each living thing has it's own set of displacements. The universe itself has an all encompassing "master displacement", a "universe wave", to which, all other displacements are added, like the sound wave signal imposed on the carrier wave in radio broadcasting, carries. Displacements of both Time and Space substance are due to a torsion motion, a twist in their substance, like the winding of a spring, which once wound, and when set free proceeds to unwind, to undo. Space and Time are related much like the front and back of a coin, they move together, displace together. **"Strain forces",** in substance, store the stress forces that caused the displacement, as the displaced substance becomes un-displaced, the stored strain forces convert back into the ability to cause stress giving substance "velocity", again.

Torque - An effort, a force that tends to produce rotation.

Torsion - The reactive torque that an elastic solid exerts, by reason of being under torsion. also the act of twisting.

Traction - To pull or tow.

Tree of Life - or The Tree of the World, it's two fruits are, eternal life through rebirth, (**reproduction**), and the second fruit is knowledge, (knowledge is knowing from **experience**, also termed enlightenment).
These are the axis about which the world of life turns, In myths of the middle east, **The tree of life, is in "The Garden of Innocence"**, The garden named Eden, God's garden.

The tree of life has been illustrated on a Painted Elamite bowl of (226-641 A.D.) as a date palm's with two fruits representing the two basic things about which life turns, knowledge and re-birth. In Parts of Asia, Pakistan, northern Africa and Egypt, the date palm, of which some 30 varieties exist. Some of these varieties are: The Deglet Noor from Algeria. The Medjool from Morocco, and three from Iraq, the Khadrawy, Zahdi, and Halaway. Also there is "The Phoenix palm", the dactylifer variety of the date palm, from Egypt. This last is often the date of commerce. It grows from 60 to 80 feet in height, has palmate leaves, who's fronds can be 10 to 20 feet long, it's fruits hang in pendulous bunches with as many as 1,000 fruits (dates), per bunch. These fruits when dried are up to half sugar. Such a tree can live up to 150 years. There are female and male palm trees, the females bear the fruit.

The fruit of Enlightenment, Enlightenment is defined as being supplied with useful information, or experience, knowledge.

The fruit of Immortality, Immortality **is** defined as, being exempt from death. This can be seen as eternal re-birth, eternal regeneration.

When snakes are shown entwined on a pole or tree trunk as in the caduceus medical symbol, they are of male and female gender, Their association with the tree of life is seen as an example of eternal life, by regeneration. The male snake and female snake entwined are matting.

The snake **originally had nothing to do with any idea of temptation, evil, or guilt.** That twisted interpretation was added at a later date by others, for their own purposes. If you have ever seen baby Western ribbon snakes (Thamnophis sauritus proximus, a slender garter snake), they are the nicest little fellows. A person in a friendly mood might name the two snakes on the caduceus medical symbol, Adam and Eve.

The Underworld - is "under heaven". T**he Underworld, is the complementary opposite of Heaven.** The world between these two extreme worlds is Earth, which exposes it's inhabitants to both "Heaven" and it's opposite "The Underworld", (termed by some hell). In the world of earth one experiences extreme ideas, extreme physical conditions, extreme treatment, and extreme expectations of others, such treatment history has shown can be outrageously unfair. Political systems, academic systems, organized religions all have been guilty of imposing such treatment on their adherents and on others, to "get their way". I believe **most men**, when not pressured to do otherwise, just want to live and let live.

Universe - The Cosmic Ocean, it's Substance and Energy. Universe is the totality of all that exists, whether as actuality or as potentiality. The Cosmic Ocean, the universe, has three basic components, which are Primordial **Substance**, Primordial **Energy** (an ability, a restlessness that animates substance). Nature's **Forces, and Counter forces,** let substance and non-substance (energy), interact. "Cosmos" and "Chaos", are the two aspects or extreme states of the Cosmic Ocean, they exist in varying portions.

Universe Waves - Great waves of displacement that cover vast distances, sloshing back and forth across the cosmic ocean. Waves of shorter wavelengths and faster frequencies, ride on these "Universe Waves". A universe wave may progress for millennia across the universe before reaching the boundary (if there is a boundary), of the universe, and reflecting off of that boundary. Universe waves, when reflected may set-up standing waves. Could the volume of the cosmic ocean be deduced from the wavelength of these standing wave swells? Have standing waves of any type ever been looked for and found in interstellar space? They may have extremely great wavelength, one may have an almost infinite wavelength, a sort of fundamental note, if we see the matter musically.

Velocity - Velocity, is speed with a definite direction of travel. Example: Meters per second with a compass direction indicated, or direction indicated by use of a vector diagram.

In the matrix diagram and in the text, I see velocity as an on-going state of motion I see it as a directed effort to bring change, as Action. Acceleration can be seen as a Re-Action to velocity's efforts. (see Acceleration in glossary).

Ideas, chemical reactions, even money can be said to have "velocity". This last sort of velocity would be the rate at which a Euro or a dollar changes hands. You buy something with your dollar, then the person you bought from now has it, then they spend it and so on. The velocity of the dollar would be how many times it changes hands in a year. Interestingly each time the dollar changes hands it becomes someone's income and the government taxes it. At years end the government may make more than a dollar in taxes on that one dollar.

The Void - Is the cosmic ocean in a state of chaos, before created things were formed. The void can be seen as an elastic solid in a state of restless, aimless, agitation. Here are brief attempts at describing the Void.
In the Rig Veda, Mandala X-129. The void was a formless substance in motion with no rigidity. In The Laws of Manu "**The Void" was "indefinable as if immersed in sleep".**

W and Z particles - Massless fictional particles to carry the nuclear weak force.
The Waking Conscious - the part of mind that perceives reality as immediate sensations streaming into it, and responds to them as needed to achieve a goal.

The Waters - The substance of the cosmic ocean. "The waters" of biblical Genesis. The uninformed darkness of The Rig Veda which states "In the beginning darkness existed enveloped in darkness. All this was indistinguishable water".

The Way - The coarse of action. The manner, method or style in which something is done. Example: She selected the way to proceed.

The weak force - is associated with neutrino interactions.

Will - Will is the power of the mind, to bring change, to compare, select, and to choose, the power to exert a directed effort to satisfy a sensed need or chosen goal. Will is potentiality as reason, the potential union of complementary opposites. Will is actuality as discrimination, judgment, decision, like an "on-off", light switch, it either holds apart or lets the flow of current continue "as is".

"Will Power", is the power to select, to choose, to initiate effort to satisfy a need, wish, desire, or intent. Example: "**I will this**", and this, to be done, or as the Egyptian pharaoh put the matter, "So it is written, so it shall be done." It was the pharaoh's "will", The power of "The Self", to control, to regulate.

From a physics point of view, "Will" (energy), has two aspects they are: **Dynamis**, (potential energy), the power to bring together, the "holding of something", this can be a "held state", or a "held opinion", this aspect reacts to change and therefore is of passive nature, it has the potential of satisfying possible goals, it has the quality of Tamas. **The other aspect of energy is, Energeia**, (kinetic energy), the power to repel, hold apart**.** This is also the power to initiate action. "Will Power", is also decision, discrimination, judgment, the act of separating by decision, this aspect has the quality "Rajas". A third aspect of will is to engage in harmony, "getting along", to the mutual benefit of the two complementary opposite aspects, this has the quality of Sattva.

Please Note: all three of these aspects of energy, or the Will, are in the vertical energy column of the matrix diagram. The Tamas quality, is in the top square, Rajas quality is in the bottom square, and the Sattva quality is in the middle square between the other two, a sort of diplomatic position.

Wisdom - "The way". The way of sharing simultaneous existence, by alternating as necessary the two complementary opposite components of a duality, to bring about a "Dynamic balance, a "Dynamic Equality", over a period of time, so each of two complementary opposites, can share existence equally, each has it's chance "To rule the roost, one does not dictate what is to be done at all times and on all things". Wisdom has the quality of "Sattva", harmony, a mutual co-operation. To resolve the tension between, between the Potential and the Actual, a "means" is needed to bring about "Dynamic Balance", or "Dynamic Equality", a measure of balance over a period of time, that "means", is a cycle that alternates two extreme aspects repeatedly. To be clear the three basic components are: Substance, Non-Substance, and the link between them "Force". The continuousness of substance carries the forces of nature. The "Individuated Self", strains to achieve a Dynamic Balance", between, the self we believe we are, and, the self we believe we are not. Emotion is the mind's way of expressing the strains of striving to achieve "dynamic balance".

Work - Work done equals effort force applied, times distance it moves an object against a constant resistance. Work done is also the force to stop the motion of a moving object over a distance. The energy expended stopping the object equals the energy expended to move it the same distance. **Caution**: You and I know that, pushing on an object and being unable to move it, is "work", energy was expended in the effort, however in a classical physics problem, the view taken is: **An effort force that produces no motion, does no work.**

Work done can be expressed in gram centimeters, kilogram meters using the metric system , or in foot pounds in non metric system.
Example: 5 pounds of force applied, object moved four feet, work done is, 5 x 4 = 20 foot pounds of work.

World Soul - of which all individual souls are a part. The soul of each thing that lives is a part of the World Soul, the soul of the whole cosmic ocean. It is by the "connectedness" of primordial substance, that one part of the universe "Feels" the forces and strains in adjacent parts. The universe itself can be seen as having a soul, a consciousness, a "Universal Soul". See also, "The Atman" in this glossary.

"Worlds" of Existence, or **"Planes of Existence".** Metaphysical Worlds. these could also be termed "Levels of Reality or "Planes of Being". There are six "worlds of Existence" in myth and ancient writings, they go by various names. Interpreting the six names of Aryan Hindu tradition as they are defined in this book, they are :Atma, Buddhi, Kama, Sthula, Manas, and Paran-Atma. Definitions of each are in this glossary.

Yahweh - The God of Israel. This may be only a part of a complete name which would be "He who brings into being whatever comes into being ", That which could do this would be "The Creative Power".

Yin (the passive) **and Yang** (the active) - Antiochus of Ascalon, Greek philosopher (born ? - died 68 B.C.) taught there are two natures, the passive and the active. As concerns the Yin - Yang doctrine, it teaches that all things, are the product of two forces, or principles, the Yin, (negative, passive, weak), and the Yang (positive, active, and strong). A chief representative of the Yin - Yang doctrine was Ysou Yen (born 340 B.C. - died 260 ? B.C.). who believed that all things, all history, and all nature, moved in cycles.

Note: The term "man", in this book generally means a person of either gender.

A Few Comments on the Author's Life

Firstly I have not earned my living as an academic, or being engaged in theoretical cosmology, or physics, so automatically I am classed as an "Amateur". Interest in the subjects you find in this book drove me into their investigation, I then wrote down my findings and conclusions. Learning to draw helps one visualize some aspects of a subject. In my case, for a brief time, when I was seven or eight, for a couple of hours, one day a week, I entered a unique "Art School", it's front door was locked and one rang a door bell to be permitted to enter. It was a Convent. The art class was taught by a nun in her black and white dress uniform. While learning to put the image on paper, I found there was "**method**" to drawing, I learned some of the characteristics of paper, the materials, and tools of the art, learned to see shapes, and manipulate colors. Eventually I noticed an elderly nun, in her sixties at the back of the classroom, quietly painting on canvas with oil based paints, She was doing a scene of an English thatched roof country cottage, with a marvelous flower garden out front, the sort of scene English watercolorist

Helen Allingham (born 1848-died 1926) did so wonderfully. Over sixty of her pictures in color can be seen in the book **"The Cottage Homes of England"**, revised edition copyright 1991 reprinted 1992, published by Bracken Books, ISBN #1-85170766-2.

The old Nun's work had shown me, what was possible with technique, a canvas, a few brushes and little tubes of colored pigment. As for my own art efforts in the class, they were confined to still life, the use of sticks of charcoal, and colored French chalk. My art education was cut short however just as I was learning to draw the human head, the rules of the order, and the facts of life had caught-up with me, it was farewell to the convent. I alas would not learn to draw young nude nuns, or even draw old nude nuns. As for my "higher education", it was spent at a school often reserved for those that find college is not in their future, one of the teachers related he had relieved one of his students of a pistol some months back, that teacher kept a Billy Club (a truncheon) on his desk, made for him in the school's fine facilities. Not being air conditioned, the classroom had a memorable atmosphere on hot humid summer days, one of the lads put it succinctly "This place smells like a hamburger joint". Eventually my education moved on to the U.S. Army. Curiously at the time I wondered, what motivated the lads to "Join-Up". So I proceeded as best I could to satisfy that gapping hole in my education. One day I and some six or seven other lads were being transported in fatigue work uniforms to an unknown (to us) location. We were riding in a closed van (a three fourth ton panel truck) with no windows in the passenger section, a lighted interior, and a long wood bench along each inside side wall, as we were bumping along I leaned forward and asked the fellow across the isle how he got in the army and he replied "Well I was on drugs and the judge gave me the choice, the army or jail". I waited a bit and then asked the lad by my right side, how he got in the army. He gave a quite straight forward explanation "Well some fellows and I were out riding on a Saturday night, the car we were riding in did not belong to any of us, and the judge gave him the choice, jail or the army." What do they popularly call the activity he was engaged in? Joy riding in a stolen vehicle? After that I did not ask any of the others I was riding with, how they got in the army, as I did not want to embarrass them. I was like **Arlo Guthrie** on Group 'W' Bench in his song **"Alice's Restaurant"** (1966-1967) only my Group 'W' Bench, was in the windowless back of a van rolling down a road on an army base.

After surviving a unique army experience, which included, meeting a fellow of Cuban decent that had served in seven different armies, yes seven, including both the French and the Spanish Foreign Legions. He wrote a manuscript for a book about his true life experiences, after talking with him I got the impression it was an honest account, it must have been a real doozey. He had submitted the manuscript to the company commander for approval. As I understand the matter, in the U.S. military, anything written for publication **must** have the company commander's approval. The experienced soldier was near the end of his military carrier and understandably rather "ticked" that he could not tell how seven armies of the world really worked, the company commander **denied approval for him to publish**. In the land of the free, or outside the land of the free, officials seldom like an honest account of the real world they are helping create and maintain.

After the military experience, I got down to earning a living. While engaged in that activity, I was officially requested to **Exorcize an Evil Spirit** from a house it was inhabiting. A rather unusual and unique request I thought, ah! you laugh, but this was serious, I had never done an exorcism before, and the occupant was truly afraid, their happiness and sense of security were at stake. I did not know what I would face when I got there, for my part I would definitely rather be spared encountering evil spirit groans, and sporadic shrieks of nightmarish orgiastic delight. To my relief at the scene of the feared spirit all was quiet except for the excited and worried resident describing the nature of the evil's presence "It's here!", she explained, and sure enough as described, it was real for I soon heard from out of the dim beyond and silence, a soft creepy "Ting, Ting, Ting" like the sound of a cymbal announcing the evil spirit's presence.
I know this will sound like I am "blowing my own horn" but to the joy of all parties concerned, mine included, my first and only exorcism was a complete and total success. In the future I would only have to face the sporadic groans and shrieks of my employers.
As concerns the writing process, I have taken the advise of a more experience author, the advise was, never discus a work in progress with others. Secondly, It was advised, **if practical**, to put my work aside for a while, then re-read it, **to see if it still makes sense**, and if spelling and any dates in the work are correct?

Those in academia need not have concern for their positions either in the classroom or when using the

Kama Sutra. I am not a graduate of Oxford, Harrow, Cal Tech, West Point, or a member of the Royal Society, that was not in my stars. I am a product of my times, not a product of an institution, still the theory expounded in this and two of my previous books **"The Man Who Saw Space and Time"** and **"I Am Universe",** I believe is reasonable, exhibits common sense, and gives stepping stones for others to use to advance science along. If this book is yours use it like a tool, mark it up with your notes, comments, corrections, conclusions, and ideas for an improved account of reality.

On more than one occasion I have been asked "Now that you are retired what do you do all day?" Well I am still learning and when I believe someone may have an answer to a lingering question, and there always seems to be, a lingering question, I ask if they can answer it. I read, reading can be a relaxing pastime, some stories tell about the countries in which they take place, mysteries such as author Michael Dibdin's Italian detective Zen Mysteries, or author Tarquin Hall's tales of India's most private investigator Vish Puri., or the spy stories of authors Manning Coles (Adelaide Manning and Cyril Coles), Cyril Coles was an actual British spy during WW 1, and possible WW 11. As has been commented about spies in general, "does a spy ever totally quit?" The Russians knew this all to well, and when Kim Philby their long time double agent, fled capture by fleeing into Russia, the Russians watched him like a hawk, overseeing his life constantly, never letting him out of Soviet block countries. Evergreen magazine (April 1968, review #53) had an article titled "My

Silent War" by Kim Philby) about Philby at the time, complete with photos of him and his wife the former Melinda Maclean in Moscow. But enough of spies, in all seriousness, in my golden years between doctor visits, I find, myself recalling friends, which if I do it, "to well", a great sadness threatens to overwhelm me, and a wish to relive being with them. I had lived with a big black cat (not fat, just respectable big), for some twenty years, I considered him a friend, and a privilege to be with, and when he died, a part of me died with him. Now how he affected my life lives on in my deeds. I would say considering this, **Blackie the Cat, is my coauthor**. Looking into the subjects of this book and writing about them has been an unexpected adventure.

Sincerely,
E. R. Margis

P.S. India's **Mohandas Karamchand Gandhi** thought that in doing one's daily tasks. "You must be the change you wish to see in the world".

Index

Persons mentioned in this book

Allingham, Hellen - English, (1848 - 1926) one of England's finest outdoor scene artists.

Antiochus of Ascalon - Greek philosopher, died about 68 B.C.

Erasmus Bartholinus - (1625-1698, Danish physicist)

Ludwig Van Beethoven, 1770-1827, German Composer.

Louis-Victor Pierre Raymond seventh Duke de Broglie - (1892-1987), French physicist.

Agatha Cristie - English authoress (1890-1976) prolific writer of mystery stories, they have been translated into different languages for the enjoyment of people around the world.

Confucius - Chinese philosopher. Grand master K'ung Fu-tzu. K'ung his family name, the tzu part means scholar or philosopher. (771 - 1123 B.C.)

Christian J. Doppler - (Austrian physicist 1803-1953).

Mohandas Karamchand Gandhi - (1869-1948, India's architect of freedom from British Imperialism by non-violent means.) Oh the word "Mahatma" is often added to Gandhi's name, it means "great soul".

Marvin Gaye - (1939-1984) American singer song writer.

Guthrie, Arlo, - American (1947 - ______), folk singer, song writer. Musician,

Heraclitus - (580c-540 B.C.) Greek philosopher.

Hermann Ludwig Ferdinand von Helmholtz - (1821-1894), German philosopher and man of science, the Helmholtz sound wave resonator is named after him.

Ursula K. Le Guin, - (Oct. 21, 1929- _______),American, author of "The Lathe of Heaven", a science fiction masterpiece.

Joseph Joubert - (1754-1824), French, philosophical thinker.

Martin Luther King Jr. - (1929-1968), American civil rights leader, who gave voice to the dream of much of mankind, to have equality and be free.

James Ephraim Lovelock - (1919 - ______), British

Sir Oliver Joseph Lodge - English physicist, (1851-1940).

Lynn Margulis (1938-2011) American? Biologist.

James Clerk Maxwell (1831-1879, Scottish physicist)

Dmitri Ivanovich Mendeleeff - or Mendelevev, (1834-1907), Russian Chemist, developer of a periodic table of the chemical elements. This table shows the relationships of the elements physical properties. by their position in the table.

Alexander A. Milne - English writer (1882-1956) author of "The Tales of Winnie the Pooh". At seventy plus years I still enjoy these adventures of a young boy, Christopher Robin and his friends, Pooh Bear, Eeyore the Donkey, Tiger, Kangaroo, Rabbit, Owl, and Piglet

Onnes, - (Kamerleigh Onnes, Heike), (Dutch Physicist (1853-1926). A master of experimental low temperature physics, among other accomplishments he liquefied Helium in 1908, he founded and developed the cryogenic laboratory at Leiden.

Max Karl Ernst Planck - (1858-1947, German theoretical physicist)

Philby Kim - (Harold Adrian Russell (Kim) Philby) (born 1912-died 1988), British). Philby was a master counterspy, possibly a triple agent, he was an undercover Soviet Union counter intelligence officer for more than thirty years (the Soviets even issued a postal stamp in 1990 with his picture on it). The British almost knighted him. The American Counter Intelligence Agency (CIA), got a spying lesson from him.

Shakespeare - (1564-1616, English playwright) He was a keen observer of the human condition. A few of the expressions and words he used, today are not common and need some explanation or defining, but are as true now as when written.

Sledge, Percy - (1940-______), American Singer of soul songs.

Swedenborg Emanuel - (1688-1772), Swedish scientist, philosopher, theologian.

Unas - **Pharaoh** in the 5th dynasty of the old kingdom about 2350 B.C.

Sir Edmund Whittaker - British, pure and applied mathematics, and theoretical physicist, teacher lecturer. (1873-1956).

Thomas Young - (1773-1829 English, amateur physicist)

Helpful Imaginary Persons.

ROY. G. BIV - Each letter represents a color in the visible spectrum in their order of occurrence, **R**ed, **O**range, **Y**ellow, **G**reen, **B**lue, **I**ndigo, and **V**iolet.

P.C.S. MAZINTAL - Used as an easy guide to electrode potential and, corrosion activity of selected metal chemical elements. Those elements at the beginning of the name **P.C.S. Mazintal**, have **higher positive electrode potential** and chemically form oxides eagerly, where those at the end of the name oxidize less readily. The letters used to make-up the name are shown here in larger bold type, along with the chemical elements they represent spelled out. **P**otassium (K), **C**alcium (Ca), **S**odium (Na), **M**agnesium (Mg), **A**luminum (Al), **Z**inc (Zn), **I**ron (Fe), **N**ickel (Ni), **T**in (Sn), Silver (**A**g), **L**ead (Pb).

The (potential) voltage range for the listed elements, runs from **+2.9v** for Potassium down to +0.44v for Iron, and further down to **- minus 0.79v** for silver (that is a minus 0.79v), Lead is just a touch more electronegative at minus 0.8v (that is minus 0.8v).

If **Hydrogen** would have been listed here it would have a zero potential of **0.0000v**. Hydrogen though is not a solid metal at room temperature such as copper, Hydrogen can however behave as a metal under some conditions, interestingly, liquid hydrogen, when subjected to two million atmospheres pressure will conduct electricity like a metal.

Hydrogen is the first element in the periodic table and the most unique.

You may wish to "look-up", a fact, or specific subject, to avoid re-reading a good part of the book, the index below, lets you quickly find many subject page number(s).

Index of Subjects

Index of Subjects, continued

Index of Subjects, continued

Index of Subjects, continued

Index of Subjects, continued

Notes

Notes

Professor

You have an opportunity to get your students "thinking outside the box", to think creatively, order each a "gift", copy of "The Heart of Reality", to help prepare them for a better future, a future in which they can judge the value of new ideas for themselves.

Also, for those who "Gots to know"

A pristine copy of the authors works can be yours. See the simplicity from which arises complexity, know the grandeur of the Cosmic Ocean, and the mystery of the natural world.

Ordering Information, for this book **"The Heart of Reality"**, or the following books by the same author with more background information on the subject, **"The Man Who Saw Space and Time"**, or **"I Am Universe"**, these can be obtained in the U.S.A. or Canada by one of the following ways.

1) By purchasing from your local book store.

2) By calling Infinity Publishing at"

1 - (877) BUY BOOK

1 - (877) - 289-2665

3) By visiting the web site www.BuyBooksOnTheWeb.com

4) or from - www.Amazon.com

Made in the USA
Monee, IL
04 November 2026

67382843R00184